Earth-Friendly Living: The Path to Sustainable Happiness

Embrace Eco-Conscious Habits for a Better Tomorrow

Isabella Coleman

© Copyright 2024 - All rights reserved.

The content contained within this book may not be reproduced, duplicated or transmitted without direct written permission from the author or the publisher.

Under no circumstances will any blame or legal responsibility be held against the publisher, or author, for any damages, reparation, or monetary loss due to the information contained within this book, either directly or indirectly.

Legal Notice:

This book is copyright protected. It is only for personal use. You cannot amend, distribute, sell, use, quote or paraphrase any part, or the content within this book, without the consent of the author or publisher.

Disclaimer Notice:

Please note the information contained within this document is for educational and entertainment purposes only. All effort has been executed to present accurate, up to date, reliable, complete information. No warranties of any kind are declared or implied. Readers acknowledge that the author is not engaging in the rendering of legal, financial, medical or professional advice. The content within this book has been derived from various sources. Please consult a licensed professional before attempting any techniques outlined in this book.

By reading this document, the reader agrees that under no circumstances is the author responsible for any losses, direct or indirect, that are incurred as a result of the use of information contained within this document, including, but not limited to, errors, omissions, or inaccuracies.

Table of Contents

INTRODUCTION 5

CHAPTER I: Understanding Sustainability 7

The Concept of Sustainability 7

The Three Pillars of Sustainability 10

The Global Impact of Unsustainable Practices 13

How Individual Actions Matter 17

CHAPTER II: Sustainable Home Practices 21

Energy Efficiency 21

Water Conservation 25

Eco-Friendly Home Products 28

Waste Reduction 31

CHAPTER III: Sustainable Food Choices 34

The Environmental Impact of Food Production 34

Adopting a Plant-Based Diet 39

Supporting Local and Organic Agriculture 43

Reducing Food Waste 50

CHAPTER IV: Eco-Conscious Transportation 56

The Environmental Cost of Conventional Transportation 56

Green Transportation Options 63

Car-Free Living 69

Tips for Reducing Transportation Footprint 78

CHAPTER V: Sustainable Fashion 86

The Environmental Impact of the Fashion Industry 86

Building a Sustainable Wardrobe 88

Thrift Shopping and Upcycling 90

Supporting Ethical Brands 93

CHAPTER VI: Eco-Friendly Parenting....................................... 96

Raising Children with Sustainable Values 96

Eco-Friendly Baby Products.. 98

Sustainable Education 101

CHAPTER VII: Green Workplaces 104

Creating Sustainable Office Environments 104

Telecommuting and Remote Work 109

Promoting Sustainability in Business 115

Encouraging Eco-Conscious Practices Among Employees ... 122

CHAPTER VIII: Sustainable Gardening and Landscaping 129

The Benefits of Sustainable Gardening 129

Techniques for Eco-Friendly Gardening 137

CONCLUSION 142

INTRODUCTION

"Earth-Friendly Living: The Path to Sustainable Happiness - Embrace Eco-Conscious Habits for a Better Tomorrow" is a groundbreaking e-book that guides cultivating a lifestyle that is both environmentally sustainable and personally fulfilling. In today's world, where concerns about climate change, resource depletion, and environmental degradation are increasingly urgent, this e-book offers practical strategies and insights for individuals seeking to impact the planet while enhancing their well-being positively.

The e-book begins by acknowledging the interconnectedness of human well-being and the planet's health. It highlights the growing recognition that our actions as individuals and as a society have profound implications for the natural world and future generations. By embracing eco-conscious habits and adopting a more sustainable way of living, readers can reduce their environmental footprint and experience a more profound sense of fulfilment and connection to the world around them.

The introduction sets the stage by framing the central themes of the e-book: the importance of sustainability, the relationship between personal happiness and environmental stewardship, and the practical steps individuals can take to align their lifestyles with these principles. It emphasizes that Earth-friendly living is not about sacrifice or deprivation but rather about making choices that promote personal and planetary well-being. By reframing sustainability as a path to greater happiness and fulfilment, the e-book seeks to inspire and empower readers to embark on their journey towards a more sustainable and fulfilling life.

Moreover, the introduction outlines the structure of the e-book, providing readers with an overview of the subjects that will be addressed in the following chapters. From practical tips for reducing waste and conserving energy to insights into the psychological benefits of nature connection and mindful consumption, each chapter offers valuable information and actionable strategies for living more sustainably. By organizing the content in a clear and accessible manner, the introduction sets the stage for readers to dive into the e-book with enthusiasm and confidence.

Overall, the introduction to "Earth-Friendly Living: The Path to Sustainable Happiness" compellingly invites readers to explore the transformative power of eco-conscious living. It establishes the foundational principles of sustainability and happiness, lays out the structure of the e-book, and prepares readers to embark on a journey towards a more fulfilling and sustainable way of life. Through its thoughtful insights and practical guidance, the e-book allows readers to make a positive difference in the world while enhancing their well-being.

CHAPTER I

Understanding Sustainability

The Concept of Sustainability

The idea of sustainability has changed dramatically over time, becoming increasingly popular as nations realize how vital it is to address global environmental, social, and economic issues. This affects not just our neighbourhood but also our globe as a whole. The capacity to fulfil demands without jeopardizing the future generations' ability to meet their own is fundamental to sustainability. This concept, made public by the Brundtland Commission in 1987, highlights the significance of long-term planning and comprehensive approaches to resource management. It also captures the multidisciplinary and intergenerational character of sustainability.

The three interrelated pillars of sustainability are social, economic, and environmental. Environmental sustainability focuses on preserving and protecting natural resources, ecosystems, and biodiversity to maintain their availability and health. This involves minimizing pollution, reducing greenhouse gas emissions, conserving energy and water, and promoting sustainable land use practices. By safeguarding the natural environment, sustainability aims to maintain ecological balance, mitigate the impacts of climate change, and protect the planet's ability to support life.

Social sustainability addresses the well-being and equity of human societies, encompassing factors such as social justice, human rights, and community resilience. It emphasizes the importance of ensuring all individuals have access to food, clean water, healthcare, education, and shelter. Social sustainability fosters inclusive and diverse communities, promotes equality and social cohesion, and addresses systemic inequalities and injustices. By prioritizing social well-being, sustainability seeks to create a world where everyone can thrive and fulfil their potential, regardless of race, gender, socioeconomic status, or other factors.

Economic sustainability focuses on creating systems and practices that support long-term prosperity and prosperity for all. It emphasizes the importance of balancing social progress, environmental preservation, and economic growth equality to ensure that financial activities contribute to the well-being of both present and future generations. This involves promoting sustainable business practices, investing in green technologies and infrastructure, and adopting economic models that prioritize people and the planet over profit. By embracing economic sustainability, societies can build resilient economies that generate prosperity while minimizing negative environmental and societal impacts.

In addition to these three pillars, sustainability also encompasses the concept of intergenerational equity, which emphasizes the responsibility of current generations to preserve and protect natural resources and ecosystems for the benefit of future generations. This principle recognizes that the choices and actions we take today will have long-lasting consequences for future inhabitants of the planet. By implementing sustainable practices and managing resources appropriately, we can ensure that the earth is resilient, healthy, and prosperous for future generations.

Sustainability is not only a theoretical idea but also a guiding notion that influences activity and decision-making at all scales, from personal preferences to international laws. We have a part to play in this, and what we do can significantly impact. It necessitates a fundamental change in perspective and conduct, questioning established growth and development paradigms and prioritizing immediate rewards above long-term sustainability. Achieving sustainability requires collaboration and cooperation across sectors and disciplines and innovative solutions that address complex and interconnected challenges.

One key challenge in advancing sustainability is balancing competing interests and priorities. For example, economic growth may be seen as incompatible with environmental protection, and social equity may be overlooked in favor of profit maximization. However, sustainable development seeks to reconcile these tensions by promoting integrated approaches considering the interdependencies between environmental, social, and economic factors. By recognizing the synergies and trade-offs between these pillars, societies can develop more holistic and effective strategies for achieving sustainability.

Addressing urgent global challenges like resource depletion, biodiversity loss, and climate change quickly presents another difficulty. These urgent concerns need our attention and action rather than being far-off threats. The Sustainable Development Goals (SDGs) offer a structure for tackling these problems and directing endeavours toward a more sustainable future. Adopted by the United Nations in 2015, the SDGs encompass various targets and indicators related to poverty eradication, environmental conservation, social inclusion, and economic development. Countries and organizations can contribute to global efforts to achieve sustainability by aligning with the SDGs and working towards their implementation.

Education and awareness are crucial for advancing sustainability and empowering individuals and communities to take action and make informed choices. Education encourages behaviour change and the development of a sustainable culture by raising awareness of sustainability challenges and highlighting the advantages of sustainable living. This includes teaching environmental literacy, promoting eco-friendly practices, and integrating sustainability into curricula at all levels of education. Public awareness campaigns, advocacy efforts, and community engagement initiatives can raise awareness about sustainability issues and mobilize support for action.

In conclusion, sustainability is a multifaceted concept encompassing environmental, social, and economic dimensions. Sustainable development necessitates integrated strategies that balance social justice, environmental preservation, and economic growth. Through adopting sustainable principles and cooperative efforts towards shared objectives, communities can establish a more resilient, just, and prosperous future for everybody.

The Three Pillars of Sustainability

Sustainability is often depicted as resting on three interconnected pillars: environmental, social, and economic. These foundations provide a framework for comprehending the complex interactions between human society and the natural world and the importance of balancing environmental protection, social equity, and economic prosperity for the well-being of present and future generations.

The environmental pillar of sustainability focuses on safeguarding natural resources, preserving biodiversity, and mitigating environmental degradation. It recognizes the finite nature of Earth's resources and the importance of maintaining ecological balance for the health and resilience of ecosystems. Environmental sustainability encompasses many practices and principles, including reducing pollution, conserving energy and water, protecting habitats, and promoting sustainable land use. By prioritizing environmental protection, sustainability ensures that ecosystems can continue providing essential services, such as clean air, water, and food while supporting the diverse species that inhabit the planet.

The social pillar of sustainability addresses the well-being and equity of human societies, emphasizing the importance of fostering inclusive and thriving communities. Social sustainability encompasses social justice, human rights, access to basic needs and services, and community resilience. It recognizes the interconnectedness of human well-being and environmental health, acknowledging that ecological degradation often disproportionately affects marginalized communities. Social sustainability seeks to promote equity, diversity, and social cohesion, ensuring that all individuals have the opportunity to lead fulfilling and dignified lives. This includes addressing systemic inequalities, promoting access to education and healthcare, supporting vulnerable populations, and building strong social networks that can provide support during times of need.

Creating structures and procedures that promote long-term prosperity while minimizing harmful consequences on the environment and society is the emphasis of the economic pillar of sustainability. To ensure that financial activities contribute to the well-being of both present and future generations, it acknowledges the significance of finding a balance between social progress, environmental preservation, and economic growth equality. Promoting sustainable corporate practices, investing in environmentally friendly infrastructure and technologies, and implementing financial models that put people and the environment before profit are all part of economic sustainability. This includes encouraging ethical labour practices, encouraging innovation and entrepreneurship that support sustainable development, and incorporating social and environmental factors into corporate decision- making.

These three pillars support and strengthen the others; they are interrelated and mutually reinforcing. For example, environmental protection is essential for safeguarding the natural resources and ecosystems on which social and economic systems depend. Likewise, social equity is necessary for ensuring that the benefits of economic development are shared equitably and that vulnerable populations are not left behind. Economic prosperity, meanwhile, provides the resources and opportunities needed to invest in environmental conservation and social development.

Achieving sustainability is not just a task for governments or corporations, but it's a responsibility that each one of us shares. A comprehensive strategy needs to be developed that considers the interactions between environmental, social, and economic factors and the trade-offs and synergies between them. This means that our individual choices, whether it's about the products we buy, the energy we consume, or the waste we generate, can make a significant difference. By embracing the three pillars of sustainability and working towards their harmonization, we can all contribute to building resilient, equitable, and prosperous futures for ourselves and for generations to come.

The Global Impact of Unsustainable Practices

The global impact of unsustainable practices extends across environmental, social, and economic dimensions, posing significant challenges to the well-being of both current and future generations. Unsustainable practices, driven by population growth, resource depletion, and consumerism, have resulted in widespread environmental degradation, social inequality, and economic instability. It is imperative that we recognize the interconnectedness of these consequences in order to address the underlying causes of unsustainability and implement workable solutions to create a more sustainable future.

Environmental degradation is the most immediate and visible consequence of unsustainable practices. Natural processes have been disturbed, ecosystems have been degraded, and biodiversity has decreased due to human activities like overfishing, deforestation, pollution, and the burning of fossil fuels. Deforestation, driven by agricultural expansion, urbanization, and logging, destroys vital habitats and contributes to climate change by releasing carbon dioxide stored in forests. Overfishing and destructive fishing practices have depleted marine resources, threatening the health and resilience of ocean ecosystems. Pollution from industrial processes, transportation, and waste disposal contaminates air, water, and soil, posing risks to human health and ecosystem integrity.

Environmental deterioration affects the natural world, human cultures, and the economy. Climate change is one of the most significant global peace and security risks, mainly because of human-made greenhouse gas emissions. Temperature increases, altered weather patterns, and sea level rise intensify population dislocation, conflict over limited resources, and food and water shortages. Climate change disproportionately affects vulnerable populations, especially those in low-lying coastal areas, desert regions, and small island states. These communities face increasing risks of food insecurity, water scarcity, and natural disasters.

Moreover, environmental degradation undermines the capacity of ecosystems to provide essential services that support human well-being, such as clean air, water, food, and climate regulation. Loss of biodiversity reduces ecosystems' resilience and increases communities' vulnerability to environmental shocks and stresses. Degraded ecosystems can absorb and mitigate the impacts of climatic catastrophes, including storms, droughts, and floods, leading to more excellent human and economic losses. Additionally, environmental degradation contributes to the spread of infectious diseases, as habitat destruction and climate change alter the distribution of disease vectors and increase the likelihood of disease outbreaks.

Social inequality and injustice are closely intertwined with unsustainable practices, exacerbating the impacts of environmental degradation on vulnerable populations. Marginalized communities, including indigenous peoples, women, children, and people with low incomes, bear a disproportionate burden of environmental degradation and are often the least equipped to cope with its consequences. Social inequities are further exacerbated, and environmental injustices, such as the location of polluting companies near low-income communities and the disproportionate distribution of environmental dangers, perp distributional inequalities. Disenfranchised populations suffer disproportionately from a lack of access to clean water, sanitary conditions, and healthcare, which raises illness, malnutrition, and death rates.

Moreover, unsustainable practices contribute to social instability and conflict by exacerbating resource scarcity, competition, and inequity. Rivalry over diminishing natural resources, like water, arable land, and minerals, can fuel tensions and conflicts within and between communities, regions, and nations. Displacement of populations due to environmental degradation, natural disasters, and conflict exacerbates social tensions and humanitarian crises as communities struggle to meet their basic needs and adapt to changing conditions. In conflict-affected areas, environmental degradation and resource scarcity can perpetuate cycles of violence, displacement, and poverty, hindering efforts to achieve peace, stability, and sustainable development.

Furthermore, unsustainable practices undermine economic prosperity and stability by depleting natural resources, increasing production costs, and creating economic dependencies on finite resources and fossil fuels. Unsustainable use and extraction of natural resources, including minerals, timber, and fossil fuels, deplete finite reserves and degrade ecosystems, diminishing their capacity to support economic activities and provide ecosystem services. Rising production costs, driven by resource scarcity, environmental regulations, and social pressures, reduce competitiveness and profitability for businesses, particularly those relying on resource-intensive industries. Economic dependencies on finite resources and fossil fuels expose economies to price volatility, supply disruptions, and geopolitical risks, undermining long-term economic resilience and stability.

Moreover, unsustainable practices contribute to economic disparities and inequalities by concentrating wealth and power in the hands of a few while marginalizing and exploiting vulnerable populations.

Extractive industries, such as mining and logging, often operate in developing countries with weak governance structures and lax environmental regulations, leading to ecological degradation, social conflicts, and human rights abuses. Child labour, forced labour, and dangerous working conditions are a few instances of exploitative labour practices that feed the cycles of exploitation and poverty and contribute to societal injustice. Furthermore, consumerism and materialism-driven unsustainable spending patterns foster a culture of waste, debt, and overconsumption that results in unstable finances, a breakdown in social relationships, and psychological suffering.

A comprehensive and integrated strategy that tackles the underlying causes of unsustainability and encourages systemic change across environmental, social, and economic aspects is needed to address the worldwide effect of unsustainable behaviours. This entails converting to renewable energy sources, encouraging sustainable resource management and land use strategies, investing in environmentally friendly infrastructure and technology, and encouraging inclusive and equitable growth. In addition, it necessitates advocating for gender equality, human rights, and social justice, as well as strengthening marginalized groups and ensuring that sustainable development's advantages are distributed fairly. It also entails encouraging waste reduction, circular economies that limit resource use and enhance resource efficiency, and sustainable production and consumption patterns. By adopting sustainability as a guiding concept and collaborative efforts to tackle its worldwide implications, civilizations may construct a more resilient, just, and prosperous future for everybody.

How Individual Actions Matter

Individual actions are not just a drop in the ocean, they are the ripples that can shape the trajectory of sustainability and environmental stewardship. While global challenges such as climate change, biodiversity loss, and resource depletion may seem daunting, it is crucial to recognize that the collective impact of individual choices and behaviours can significantly and positively affect the environment, society, and economy. Individuals can support more significant initiatives to address environmental and social issues, promote sustainability, and build a more resilient and equitable world by incorporating sustainable habits into their daily lives.

Recognizing one's agency and duty in bringing about positive change lies at the core of individual action. Every choice, whether to consume resources or buy items, affects society and the environment. Practicing mindfulness and intentionality in our decision-making can reduce our environmental impact and optimize our beneficial contributions to sustainability. This entails trying to make decisions consistent with our beliefs and ambitions for a better world and taking into account the effects our actions will have on the environment, society, and economy.

One of the most empowering ways individuals can make a difference is through conscious consumption. People may encourage companies that prioritize sustainability by supporting them and encouraging others to follow suit by selecting goods and services that are ethically sourced, ecologically friendly, and sustainably created. This entails purchasing goods with little packaging, consuming organic and locally sourced food, and picking solid and long-lasting products that are recyclable and reusable. Furthermore, people can embrace minimalism and prioritize quality over quantity to reduce waste and conserve resources by adopting a more minimalist lifestyle.

One crucial area where individual actions can greatly impact sustainability is energy consumption. People may significantly contribute to mitigating climate change, reducing air and water pollution, and protecting natural resources by being conscious of how much energy they use, encouraging energy efficiency, and embracing renewable energy sources. This can be as easy as adopting energy-efficient appliances, disconnecting devices, turning off lights when not in use, and thinking about generating electricity for your home using solar or wind turbines. People can also support policies and initiatives that promote renewable energy and advocate for structural reforms that will make the shift to a low-carbon economy easier.

Another crucial component of personal sustainability is transportation. By choosing sustainable modes of transportation like walking, bicycling, carpooling, or using public transportation, people can enhance air quality, lessen transportation congestion, and emit fewer greenhouse gases. This benefits the environment, reduces reliance on fossil fuels, promotes physical exercise, and reduces transportation costs. People should also favor upgrades to the infrastructure supporting sustainable transportation, such as bike lanes, pedestrian walkways, and electric vehicle charging stations, to make sustainable options more accessible and easier for everyone.

Waste reduction is a fundamental component of individual sustainability. By minimizing waste generation, practicing recycling and composting, and choosing reusable alternatives to single-use items, individuals can reduce the burden on landfills, conserve resources, and protect ecosystems. This involves adopting a "zero-waste" mindset and embracing bulk shopping, meal planning to reduce food waste, and composting organic materials. Additionally, individuals can support policies and initiatives that promote waste reduction, such as bans on single-use plastics, extended producer responsibility programs, and incentives for recycling and composting.

Water conservation is another important aspect of individual sustainability, particularly in regions facing water scarcity and drought. By reducing water usage, fixing leaks, and implementing water-saving practices at home and in daily activities, individuals can conserve water resources, protect ecosystems, and ensure access to clean water. This entails installing water-saving fixtures, water-wise landscaping with greywater, and water-conscious behaviors like shutting. When cleaning your teeth or doing the dishes, turn off the faucet. People can also back laws and programs that encourage water conservation, like those that impose water prices, encourage drought-resistant landscaping, and fund water-saving devices.

Individual sustainability is also significantly influenced by food choices. Individuals can reduce greenhouse gas emissions, conserve land and water resources, and promote animal welfare by opting for plant-based foods, reducing food waste, and supporting sustainable agriculture practices. This involves embracing a more plant-centric diet, incorporating locally grown and seasonal foods into meals, and choosing organic and sustainably produced foods whenever possible. People can also fund projects encouraging sustainable farming methods like agroecology and regenerative agriculture, farmers' markets, and community-supported agriculture programs.

Individual acts can also involve advocacy, education, community involvement, and consuming decisions. Individuals can increase their effect and accelerate societal change by pushing for policy reforms, getting involved in community efforts, and spreading awareness of sustainability challenges. This can include sharing information about sustainability on social media, participating in environmental campaigns and protests, volunteering with local organizations, and engaging with policymakers to advocate for sustainable solutions. Additionally, individuals can support environmental education programs in schools, workplaces, and communities to empower others with the knowledge and skills needed to take action for sustainability.

In conclusion, individual actions matter significantly in pursuing sustainability and environmental stewardship. People can effect good change and create a more sustainable world by adopting mindful decisions regarding consumption, energy use, transportation, waste management, water conservation, and food intake. Moreover, by advocating for policy changes, raising awareness, and engaging with communities, individuals can amplify their impact and inspire others to join the collective effort toward sustainability. Ultimately, the collective power of individual actions is essential for addressing global challenges, promoting sustainability, and creating a more resilient and equitable future for all.

CHAPTER II

Sustainable Home Practices

Energy Efficiency

Energy efficiency, which has several advantages for the economy, society, and environment, is essential to sustainability. Fundamentally, energy efficiency is about utilizing less energy to do the same tasks or reach the same productivity and comfort levels. Energy efficiency contributes to decreased energy consumption and fewer greenhouse gas emissions while protecting the environment's natural resources by minimizing energy waste and optimizing energy utilization. Furthermore, energy efficiency can save people, companies, and governments costs while fostering economic competitiveness, innovation, and job development.

One of the critical advantages of energy efficiency is its contribution to mitigating climate change and reducing environmental impact. Fossil fuel combustion contributes significantly to the release of greenhouse gases, which ultimately cause global warming and climatic instability. By improving energy efficiency in buildings, transportation, industry, and appliances, we can reduce the demand for fossil fuels and lower carbon dioxide emissions and other pollutants. This helps mitigate climate change, reduces air and water pollution, improves public health, and protects ecosystems and biodiversity. Furthermore, energy efficiency can improve energy security by lowering reliance on imported fossil fuels and unstable energy markets and boosting resistance to supply disruptions and price swings.

Energy efficiency isn't just about saving the planet, it's about saving money too. It offers significant economic benefits, both at the individual and societal levels. For individuals and households, energy efficiency measures such as insulation, weatherization, and energy-efficient appliances can lead to lower energy bills and increased disposable income. By investing in energy efficiency upgrades, homeowners and renters can reduce their energy costs and improve the comfort and value of their homes. Similarly, businesses and industries can achieve cost savings through energy efficiency measures such as process optimization, equipment upgrades, and energy management systems. These savings can be reinvested in other areas of the business, such as research and development, employee wages, and expansion, leading to increased productivity and competitiveness.

At the societal level, energy efficiency can stimulate economic growth, create jobs, and foster innovation. Investments in energy efficiency projects, such as building retrofits, infrastructure upgrades, and renewable energy installations, generate demand for skilled labor and create job opportunities in construction, manufacturing, engineering, and other sectors. Moreover, energy efficiency policies and programs can spur innovation and technological development, driving the adoption of new technologies and practices that improve energy efficiency and reduce environmental impact. This innovation can lead to new industries, products, and services, increased exports and international competitiveness.

In addition, attaining sustainable development objectives like lowering poverty, enhancing health and well-being, and encouraging inclusive and equitable economic growth depend heavily on energy efficiency. It is necessary to have access to reasonably priced and dependable energy services to fulfill necessities for cooking, heating, cooling, lighting, and communication. By improving energy efficiency and expanding access to energy-efficient technologies, we can ensure everyone can access clean, reliable, and affordable energy sources, regardless of income level or geographic location. This can help alleviate energy poverty, improve living standards, and enhance the quality of life for billions of people worldwide.

In addition to its environmental and economic benefits, energy efficiency also offers social advantages, such as improved indoor air quality, increased comfort and productivity, and enhanced resilience to extreme weather events. Energy-efficient buildings, appliances, and transportation systems can help to reduce indoor air pollution, allergens, and respiratory illnesses, leading to better health outcomes for occupants. Moreover, energy-efficient buildings are often more comfortable and pleasant to live and work in, with stable temperatures, improved lighting, and reduced noise levels. This can result in lower absenteeism and employee turnover for companies and employers and greater productivity, creativity, and well-being for residents. Additionally, energy-efficient solutions like weatherization and building insulation can increase resistance to extreme weather events like heatwaves, storms, and floods by lowering energy consumption and lessening the effects of temperature changes.

Energy efficiency has several drawbacks despite its many advantages. The initial cost of energy efficiency upgrades is one of the most enormous obstacles, and it can be a significant deterrent for people, companies, and governments—particularly in developing nations or areas with little financial resources. On the other hand, many energy-saving solutions have short payback times and long-term cost savings, making them wise investments in the long run. Monetary incentives can also help defray the upfront costs of energy efficiency renovations and increase their accessibility and affordability for individuals and companies. These incentives include tax credits, grants, rebates, and low-interest loans.

Another challenge is more awareness and information about energy efficiency opportunities and benefits, particularly among consumers, businesses, and policymakers. Many people must know the energy-saving potential of simple measures such as turning off lights, insulating homes, and upgrading to energy-efficient appliances. Moreover, misconceptions about energy efficiency, such as the belief that it requires sacrificing comfort or productivity, can deter individuals and businesses from taking action. Education and outreach efforts, such as public awareness campaigns, energy audits, and energy efficiency labeling programs, can raise awareness and provide information about energy-saving opportunities and benefits.

Furthermore, policy and regulatory barriers can hinder the adoption of energy efficiency measures and technologies, such as outdated building codes, lack of energy efficiency standards, and insufficient incentives for energy efficiency investments. However, supportive policies and regulations can help to overcome these barriers and create an enabling environment for energy efficiency. This covers energy efficiency goals and incentives, financial tools like carbon pricing and emissions trading, and energy performance requirements for structures and appliances.

Governments can also encourage energy efficiency by sponsoring research and development, establishing capacity, and enacting rules about public procurement.

Energy efficiency is a worldwide issue. It's essential for tackling issues like economic inequality, energy insecurity, and climate change. The environment, economy, and society can all benefit greatly from energy efficiency, which also saves natural resources and reduces energy waste and emissions. Furthermore, energy efficiency is a practical and affordable solution that can be applied at all scales, from private residences to major corporations and governmental bodies. But in order to fully utilize energy efficiency, several obstacles must be overcome.

Water Conservation

In the current global setting, conserving water is not just a decision to be made but also a need. It is essential to sustainable resource management to guarantee that there will be clean, safe water for present and future generations. With global water scarcity reaching alarming levels, effective water conservation strategies are no longer an option, but a must. They are essential to mitigate the impacts of water shortages, preserve ecosystems, and maintain water quality. Everyone, from individual households to large industries and agricultural operations, has a role to play in conserving water and promoting responsible water use.

At the heart of water conservation efforts is the recognition of freshwater resources' finite nature and the need to use them wisely and efficiently. Freshwater is a limited and vulnerable resource, with only a tiny fraction of the Earth's water available for human use. Climate change, urbanization, industry, and population growth are putting unprecedented pressure on freshwater sources, leading to overexploitation, pollution, and depletion of water resources. By conserving water and adopting sustainable water management practices, we can ensure that water remains available for essential needs such as drinking, sanitation, agriculture, industry, and ecosystem health.

As individuals, we have the power to make a significant impact on water conservation. One of the most effective ways to do this is by reducing water waste and promoting water efficiency in our daily lives. This can be as simple as fixing dripping faucets and running toilets, and investing in water-efficient appliances and fixtures. Water conservation can also be aided by little practices like taking shorter showers, shutting off the faucet when cleaning dishes or brushing your teeth, and sweeping driveways and sidewalks with a broom rather than a hose. We can all contribute to preserving our valuable water resources by implementing these habits.

Water conservation is not just about preserving water resources, but also about ensuring sustainable food production and livelihoods. Agriculture, as the largest consumer of freshwater globally, plays a significant role in this. However, much of this water is wasted due to inefficient irrigation practices, waterlogging, and evaporation. By adopting water-saving irrigation techniques such as drip irrigation, micro-sprinklers, and precision farming technologies, farmers cannot only reduce water use but also improve crop yields and minimize water-related environmental impacts. Moreover, agroecological practices such as conservation tillage, crop rotation, and agroforestry can enhance soil health, water retention, and resilience to drought, thereby reducing the need for irrigation and promoting water conservation.

Water conservation is essential to reduce water use, minimize wastewater generation, and protect water quality. Industries use large volumes of water for various purposes, including manufacturing, cooling, and cleaning. However, much of this water is discharged as wastewater, often containing pollutants that can harm aquatic ecosystems and human health. Industries can reduce water consumption, lower operating costs, and minimize their environmental footprint by implementing water-saving technologies and practices such as water recycling, closed-loop systems, and process optimization. Additionally, industries can invest in water treatment technologies to treat and reuse wastewater, thereby conserving water resources and reducing pollution.

Furthermore, water conservation efforts are essential for protecting freshwater ecosystems and biodiversity. Healthy freshwater ecosystems, including rivers, lakes, wetlands, and aquifers, provide crucial services such as water purification, flood control, habitat for wildlife, and recreational opportunities. However, these ecosystems are under increasing pressure from pollution, habitat destruction, over-extraction, and climate change. We may contribute by lowering pollution, preserving water, protecting and restoring freshwater ecosystems, and ensuring their continued health and resilience. This includes restoring riparian buffers, reducing nutrient runoff from agriculture, and implementing sustainable water management practices in urban and rural areas.

In addition to individual and local efforts, effective water conservation requires coordinated action at the regional, national, and global levels. Governments are crucial in setting policies, regulations, and incentives to promote water conservation and sustainable water management. This includes measures such as water pricing mechanisms, water rights allocations, water quality standards, and investments in water infrastructure, research, and education. Moreover, international cooperation and partnerships are essential for addressing transboundary water issues, sharing best practices, and mobilizing resources for water conservation and sustainable development.

Education and awareness are also critical components of water conservation efforts, empowering individuals, communities, and organizations to act and make informed water-use choices. We can foster a culture of water stewardship and collective responsibility by raising awareness about the importance of water conservation, promoting water-saving technologies and practices, and providing training and capacity-building opportunities. Moreover, public engagement and participation can help mobilize support for water conservation initiatives, build social cohesion, and strengthen resilience to water-related challenges.

In conclusion, water conservation is essential for guaranteeing that both the present and future generations can access clean, safe water. By reducing water waste, promoting water efficiency, and adopting sustainable water management practices, we can conserve water resources, protect freshwater ecosystems, and promote resilience to water-related challenges. From individual actions to collective efforts, everyone has a role in conserving water and promoting responsible water use. By working together and embracing a culture of water stewardship, we can build a more sustainable and resilient future for all.

Eco-Friendly Home Products

In today's world, where environmental consciousness is a top priority, the demand for eco-friendly home products has skyrocketed. These products, designed with sustainability in mind, not only help consumers reduce their carbon footprint but also offer practical benefits. From renewable materials that are durable and long-lasting to energy-efficient appliances that lower utility bills, eco-friendly home products are a smart choice for those looking to live sustainably.

Renewable energy solutions are among the most notable categories of eco-friendly home products. For example, solar panels are a popular way to harness solar energy to create clean, renewable electricity. These panels may be installed on roofs to provide domestic electricity, which helps lessen reliance on fossil fuels and carbon emissions. Furthermore, due to developments in solar technology, these systems are now more cost-effective and efficient, giving homeowners wishing to switch to renewable energy sources a feasible alternative.

Another essential aspect of eco-friendly home products is energy-efficient appliances. From refrigerators to washing machines, many household appliances now come with energy-saving features that help conserve electricity and water. For example, Energy Star-certified appliances are designed to consume less energy during operation, leading to lower utility bills and decreased environmental impact. Further improving energy efficiency within the home are inventions like smart thermostats, which enable homeowners to manage their energy usage by altering temperature settings based on occupancy and preferences.

Beyond energy, eco-friendly home products extend to sustainable materials and construction practices. For instance, bamboo flooring has become a renewable alternative to traditional hardwood floors. Bamboo proliferates and requires minimal pesticides or fertilizers, making it an eco-friendlier choice for flooring material.

Similarly, reclaimed wood products, such as furniture and accent pieces, offer a sustainable option by repurposing wood from old buildings or structures, reducing the need for new timber and minimizing waste.

Concerns for the environment and public health are driving the transition to more environmentally friendly household cleaning products. Toxins and harsh chemicals found in many conventional cleaning products can be dangerous for the environment and human health. On the other hand, eco-friendly substitutes promote a better living environment for your family and cause less harm to ecosystems when rinsed down the drain because they contain natural and biodegradable components that are safer to use around the house.

Water conservation is also a significant factor in eco-friendly household products. Low-flow fixtures, like showerheads, toilets, and faucets, limit the flow rate without compromising functionality to help cut down on water usage. Furthermore, homeowners can lessen their dependency on municipal water supplies and preserve freshwater by gathering and maintaining rainfall for non-potable uses like toilet flushing and plant watering.

In summary, eco-friendly home goods empower users to make environmental decisions that positively impact their daily lives. These goods help create a more sustainable and environmentally friendly home environment through water conservation techniques, energy-efficient appliances, renewable energy sources, or sustainable materials. By introducing these goods into their homes, people may lessen their carbon footprint, save essential resources, and help create a better planet for coming generations.

Waste Reduction

Waste reduction has emerged as a critical component of sustainable living in response to the escalating global waste crisis. As populations grow and consumption rates increase, the amount of waste generated continues to climb, placing immense strain on landfills, ecosystems, and natural resources. In this context, efforts to minimize waste production and maximize resource efficiency have become paramount. Waste reduction encompasses a spectrum of strategies aimed at reducing waste generation at its source, diverting materials from landfills through recycling and composting, and promoting reuse and repurposing to extend the lifespan of products and materials.

At the forefront of waste reduction efforts is the principle of waste hierarchy, which prioritizes waste management strategies based on their environmental impact. At the top of the hierarchy is waste prevention, which focuses on reducing the amount of waste generated in the first place. This can be achieved through initiatives such as product design for durability and longevity and the promotion of minimalist lifestyles that prioritize experiences over material possessions. By preventing waste from being created in the first place, resources are conserved, and environmental impacts are minimized.

In addition to waste prevention, waste reduction strategies also encompass waste diversion techniques aimed at diverting materials from landfills through recycling and composting. Recycling entails repurposing garbage, preserving raw materials, and cutting down on energy used in the extraction and production. Commonly recycled materials include paper, plastics, glass, and metals, which can be sorted, processed, and reintroduced into the production cycle. In contrast, composting breaks down organic waste into nutrient-rich compost that may be used to fertilize gardens and enhance soil. Examples of these items include food scraps and yard waste. Composting prevents methane emissions, a potent greenhouse gas created during anaerobic decomposition, from entering landfills by rerouting organic waste.

Furthermore, waste reduction efforts emphasize the importance of reuse and repurposing to extend the lifespan of products and materials. Reuse involves using products or materials multiple times before discarding them, thereby delaying their entry into the waste stream. This can be achieved through initiatives such as refillable containers, reusable shopping bags, and secondhand markets that promote exchanging and reusing goods. Repurposing, on the other hand, involves creatively transforming waste into new products or materials with a different purpose or function. This can range from upcycling old clothing into new garments to repurposing glass jars into storage containers or home decor. Giving new life to discarded materials, reusing them, and repurposing them reduces the demand for new resources and minimizes waste generation.

To address the underlying causes of waste generation, waste reduction calls for immediate systemic changes at the societal level and individual initiatives. This entails enforcing laws and rules to encourage waste reduction and advance sustainable business practices. For instance, extended producer responsibility (EPR) programs hold manufacturers responsible for managing their goods' end-of-life, enabling the design of reusable and recyclable products. Similarly, landfill levies and trash disposal prohibitions promote investment in infrastructure for recycling and waste avoidance while discouraging the disposal of valuable materials in landfills.

Governments may foster an atmosphere that supports sustainable waste management practices by incorporating waste reduction principles into policy frameworks.

In conclusion, waste reduction is not just a concept, but a responsibility we all share in mitigating the environmental, social, and economic impacts of waste generation. Waste reduction, recycling, composting, reusing, and repurposing are essential practices we can all prioritize to lessen our environmental effects and contribute to building a more sustainable future. More sustainable future. Governments, businesses, communities, and individuals must work together and coordinate to implement comprehensive solutions that address the underlying causes of waste generation and advance a circular economy that seeks to achieve significant advancements by reducing waste and maximizing resource efficiency.

CHAPTER III

Sustainable Food Choices

The Environmental Impact of Food Production

The environmental impact of food production is a multifaceted issue encompassing various aspects of agriculture, livestock rearing, fishing, and food processing. This topic is paramount as the global population continues to rise, increasing the demand for food and intensifying the strain on natural resources and ecosystems. Understanding the environmental consequences of food production requires examining how different farming practices, food supply chains, and consumption patterns affect the planet.

Agriculture forms the backbone of food production and has significant environmental repercussions. One of the primary concerns is the extensive use of land for growing crops. Large-scale monoculture farming, in which enormous tracts of land are planted exclusively for one crop, destroys natural habitats to create room for agricultural fields, which results in a decline in biodiversity. Ecosystems are disturbed by this decline in biodiversity, which increases their susceptibility to pests and illnesses and frequently results in a rise in chemical pesticides and herbicides. These substances have the potential to poison waterways and soil, endangering aquatic life and decreasing soil fertility.

Another major environmental issue associated with agriculture is water use. Irrigation practices, especially in arid and semi-arid regions, place enormous pressure on freshwater resources. About 70% of freshwater withdrawals worldwide come from agriculture; in certain places, this depletion of non-renewable aquifers is caused by water extraction from these sources. Over-irrigation can also lead to soil salinization, where the accumulation of salts in the soil impairs crop growth and reduces agricultural productivity. In addition to direct water use, agriculture also impacts water quality through runoff that carries fertilizers and pesticides into rivers and lakes, causing eutrophication and dead zones where aquatic life cannot survive.

Soil degradation is another critical concern in food production. Intensive farming practices, including excessive tilling, monoculture planting, and reliance on synthetic fertilizers, can lead to soil erosion, loss of organic matter, and declining soil health. Erosion removes the nutrient-rich topsoil, making land less productive and more susceptible to desertification. Moreover, synthetic fertilizers can disrupt the natural nutrient cycles, leading to soil acidification and a decline in beneficial soil organisms essential for maintaining soil structure and fertility.

Livestock production, particularly the industrial-scale rearing of cattle, pigs, and poultry, has profound environmental impacts. One of the most critical problems is the emission of greenhouse gases. Livestock digest their food and create methane, a potent greenhouse gas, through enteric fermentation. Cattle alone account for a sizable amount of the methane emissions from livestock that contribute significantly to global warming. An additional potent greenhouse gas is nitrous oxide, produced by managing manure and the breakdown of organic matter in feedlots and pastures.

Deforestation is closely linked to livestock production, particularly in regions like the Amazon rainforest. Forests are cleared to make pastureland for grazing or to cultivate feed crops like soybeans. In addition to reducing biodiversity and the ability to sequester carbon, deforestation disturbs water cycles and exacerbates soil erosion. The conversion of forests into agricultural land is one of the primary causes of habitat degradation and the loss of species.

Fishing and aquaculture methods also have a significant negative influence on the ecosystem. Fish populations are depleted by overfishing, endangering marine biodiversity and the means of subsistence for communities that depend on fishing. Seagrass beds and coral reefs can sustain significant damage from destructive fishing techniques like bottom trawling. While seen as a solution to overfishing, Aquaculture presents its own challenges. Fish farms can lead to water pollution from feed and waste, spreading diseases to wild fish populations, and escaping non-native species that may disrupt local ecosystems.

The environmental footprint of food production extends beyond farming and livestock rearing to include food processing, packaging, and transportation. Food processing involves significant energy consumption and water, chemicals, and other resources. Food product packaging adds to the amount of plastic trash produced, which is dumped, contaminating the environment and putting wildlife in peril, either in landfills or the ocean. Food goods emit carbon emissions during transportation, significantly when delivered over great distances, which adds to global warming.

Food waste is another major issue with substantial environmental and mental impacts. A significant quantity of food is lost or wasted along the whole food supply chain, from production and processing to retail and consumption. This waste not only wastes the water, energy, and land needed to produce the food, but as it decomposes in landfills, it also emits greenhouse gases into the sky. Therefore, decreasing food waste is crucial to lessening food production's environmental harm.

A number of sustainable practices and laws are implemented to lessen the harm that food production causes to the environment. Crop rotation, agroforestry, conservation tillage, and organic farming are examples of sustainable agriculture techniques that can support soil health, water conservation, and a decrease in the need for chemical inputs. Integrated pest management (IPM) techniques, which combine biological, cultural, and chemical methods to control pests, can reduce reliance on harmful pesticides.

In livestock production, sustainable practices include improving feed efficiency, adopting rotational grazing, and implementing better manure management systems. These practices can help reduce methane emissions, enhance soil health, and protect water quality. Additionally, alternative protein sources, such as plant-based proteins and lab-grown meat, are being developed to reduce the environmental footprint of meat production.

Sustainable methods for aquaculture and fisheries are also crucial. These include limiting bycatch, preserving important habitats, enhancing feed and waste management in fish farms, and establishing catch limits based on scientific evaluations. Customers can learn more about the sustainability of seafood products from certification programs like the Aquaculture Stewardship

Council (ASC) for farmed fish and the Marine Stewardship Council (MSC) for wild-caught fish.

Promoting sustainable food production requires the active participation of policymakers. Policies that support sustainable agricultural practices, provide incentives for reducing food waste, and regulate the use of chemicals and water in farming are essential. Additionally, trade policies can impact the environmental footprint of food production by influencing where and how food is produced and transported.

Consumer choices also significantly impact the environmental footprint of food production. Choosing locally produced, seasonal, and organic foods can reduce the demand for resource-intensive farming practices and long-distance transportation. Reducing meat consumption and choosing plant-based alternatives can lower greenhouse gas emissions and decrease the pressure on land and water resources.

Awareness and education campaigns are essential to advancing sustainable food production and consumption. By informing consumers about the environmental impacts of their food choices and encouraging sustainable practices, these campaigns can drive demand for environmentally friendly products and practices.

In conclusion, addressing the complex and varied issue of the environmental impact of food production calls for an all-encompassing strategy. The environmental effects of food production can be significantly reduced by using sustainable agricultural practices, ethical animal management, sustainable fishing and aquaculture, efficient legislation, informed consumer choices, and public education. By using these tactics, we can work toward a food system that meets human needs and promotes the health of the world.

Adopting a Plant-Based Diet

Adopting a plant-based diet involves eliminating or avoiding animal products and emphasizing consuming predominantly plant-based foods and beverages, including grains, seeds, nuts, fruits, legumes, and other plant-based foods. This dietary change has attracted a lot of interest because of the possible health advantages, moral implications, and favourable environmental effects. The acceptance of plant-based diets is increasing as more people look for ways to enhance their sustainability and health. This trend is bolstered by mounting data and lobbying from environmentalists, animal rights advocates, and nutritionists.

It has been demonstrated that a plant-based diet offers many health advantages. Numerous studies suggest that eating a diet heavy in plant-based meals can reduce your chance of developing long-term conditions like obesity, type 2 diabetes, heart disease, and several cancers. One of the key factors contributing to these health benefits is the high nutrient density of plant-based foods. Fruits, vegetables, legumes, nuts, and seeds contain essential vitamins, minerals, antioxidants, and dietary fibre. These nutrients are crucial in maintaining optimal health by supporting immune function, reducing inflammation, and promoting healthy digestion. For instance, dietary fibre, abundant in whole plant foods, aids in maintaining healthy bowel movements, lowering cholesterol levels, and regulating blood sugar levels, can help prevent and manage conditions like diabetes and heart disease.

Furthermore, diets based on plants often contain less cholesterol and saturated fats, which are frequently present in animal products and are known to have a role in the development of cardiovascular disorders. Healthy plant-based fats, such as those in nuts, seeds, and avocados, can replace animal fats in diets to enhance lipid profiles and lower the risk of heart disease. Furthermore, plant-based diets usually have fewer calories and more nutrients that encourage satiety, which can aid in weight management and reduce obesity's prevalence; it significantly contributes to the likelihood of developing certain chronic illnesses.

The ethical considerations surrounding the adoption of a plant-based diet primarily revolve around animal welfare and the moral implications of consuming animal products. Factory farming, the dominant method of animal production in many parts of the world, often involves inhumane practices and living conditions for animals. These include overcrowded and unsanitary environments, routine use of antibiotics and hormones, and practices such as debeaking and tail docking without anaesthesia. People can oppose these methods and encourage the more humane treatment of animals by deciding to consume a plant-based diet. Additionally, adopting a plant-based diet aligns with the ethical principle of reducing harm, as it minimizes the need for animal slaughter and exploitation.

From an environmental perspective, adopting a plant-based diet offers substantial benefits. The food production system, particularly the livestock sector, contributes to environmental degradation. Livestock farming is responsible for significant global greenhouse gas emissions, including methane, nitrous oxide, and carbon dioxide, which contribute to climate change. Methane is a potent gas created during enteric fermentation in ruminant animals like cows. Its potential to cause global warming is many times greater than that of carbon dioxide. People can significantly minimize their carbon footprint and the consequences of climate change by reducing or giving up animal product consumption.

Moreover, raising cattle contributes significantly to deforestation, especially in areas like the Amazon rainforest, where forests are cut down for pastureland and feed crops like soybeans. Ecosystem disruption, a decrease in biodiversity, and the atmospheric release of carbon dioxide that has been stored are all results of this deforestation. On the other hand, compared to animal-based meals, plant-based diets typically demand fewer natural resources, such as water and land. Plant-based diets are more efficient than animal feed since they require less space and water to grow crops for human use. This effectiveness can safeguard biodiversity and aid in the preservation of natural environments.

Water usage is another critical environmental issue linked to food production. Animal agriculture is water-intensive, with large amounts required for drinking, feed crop irrigation, and processing. In contrast, plant-based foods typically have a lower water footprint. For instance, a pound of beef requires much more water than the same quantity of plant-based protein sources, like beans or lentils. People can reduce their impact on freshwater habitats and help preserve water supplies by adopting a plant-based diet.

Conventional livestock agriculture raises severe environmental issues related to soil health and land degradation. Overgrazing by livestock can result in soil erosion, desertification, and loss of fertility. Furthermore, the extensive use of chemical pesticides and fertilizers in the production of feed crops can contaminate water and soil, endangering wildlife and lowering the sustainability of agriculture. Plant-based diets can support long-term agricultural productivity and soil health because they emphasize varied crop production and sustainable farming methods.

The shift to a plant-based diet also affects society and the economy. Animal agriculture's industrialization has resulted in farm consolidation and a fall in family-run and small-scale operations. This change has significantly impacted rural communities, causing social unrest, job losses, and unstable economies. By supporting plant-based food systems and sustainable agriculture, individuals can contribute to revitalizing rural economies, supporting local farmers, and promoting fair labor practices.

In addition to the broad benefits of adopting a plant-based diet, there are practical considerations for the transition. Many people are concerned about meeting their nutritional needs without animal products, particularly protein, vitamin B12, iron, calcium, and omega-3 fatty acids. On the other hand, a carefully thought-out plant-based diet can supply every nutrient needed for optimal health. Legumes (including beans, lentils, and chickpeas), nuts, seeds, entire grains, and items made from soy (such as tempeh and tofu) are a few plant sources that provide protein. Since vitamin B12 is not found naturally in plant foods, supplements or fortified meals (such as plant-based milk and cereals) can be used to get it. Plant-based sources of calcium, iron, and omega-3 fatty acids are also available, and eating various plant foods can help guarantee that these nutrients are ingested in sufficient amounts.

Education and access to resources are crucial for supporting individuals in transitioning to a plant-based diet. Cooking classes, nutritional counselling, and community support groups can provide valuable information and encouragement. Additionally, the availability of plant-based foods in grocery stores, restaurants, and cafeterias makes it easier for people to make plant-based choices. Food companies and chefs increasingly offer innovative and delicious plant-based options, making the transition more accessible and enjoyable.

The adoption of a plant-based diet also has cultural and culinary dimensions. Many traditional cuisines worldwide are inherently plant-based or include a significant proportion of plant-based dishes. Embracing a plant-based diet can involve exploring and celebrating these culinary traditions, which often emphasize fresh, seasonal, and minimally processed ingredients. This exploration can enrich one's diet, introduce new flavors and textures, and foster a greater appreciation for diverse food cultures.

In conclusion, a plant-based diet offers many benefits for individual health, animal welfare, and environmental sustainability. By reducing the consumption of animal products and increasing the intake of plant-based foods, individuals can lower their risk of chronic diseases, take a stand against inhumane animal practices, and reduce their environmental footprint. Plant-based diets are easy to adopt and satisfying to follow because of the abundance of available nutritional knowledge, valuable tools, and cultural customs that support them. More people will probably choose a plant-based lifestyle as the advantages of these diets become more widely known, which will help ensure a healthier and more sustainable future for all.

Supporting Local and Organic Agriculture

It's imperative to support the organic and local agriculture component of creating a sustainable food system that benefits the environment, local economies, and community health. This practice emphasizes purchasing and consuming food grown or produced near where it is sold, cultivated without synthetic pesticides, fertilizers, genetically modified organisms (GMOs), or other harmful agricultural practices. The rise of the local and organic movement has been driven by growing consumer awareness about the environmental, health, and social impacts of conventional industrial agriculture and a desire to reconnect with food sources and support more resilient food systems.

One of the main advantages of supporting local agriculture is decreasing the distance that food travels, or "food miles," from the source of production to the consumer. Customers can reduce transportation's carbon footprint by consuming food grown nearby. This transportation reduction decreases greenhouse gas emissions and reduces the need for energy-intensive refrigeration and packaging. Fresh, locally sourced produce often requires less packaging, decreasing plastic and other waste that can contribute to environmental pollution.

Moreover, local agriculture supports biodiversity and the preservation of heirloom and native plant varieties. Small-scale local farms are more likely to grow a diverse range of crops than large industrial farms that often focus on monoculture. Crop diversity is crucial for maintaining healthy ecosystems, as it promotes soil health, reduces the spread of pests and diseases, and provides habitats for beneficial insects and wildlife. Consumers can help preserve agricultural biodiversity and promote resilient farming systems by supporting local farmers who cultivate various crops.

Supporting organic agriculture also has significant environmental benefits. Organic farming practices prioritize soil health through crop rotation, cover cropping, and applying organic compost and manure. These practices enhance soil fertility, structure, and water retention, producing more productive and sustainable agricultural systems. Furthermore, artificial fertilizers and pesticides, which may be detrimental to the environment, are not used in organic farming. Synthetic chemicals can contaminate soil and water, harm non-target species, and contribute to the decline of pollinators such as bees and butterflies. Consumers choose organic products to support farming methods that harmonize with natural ecosystems and protect biodiversity.

Another critical aspect of supporting local and organic agriculture is improving food security and community resilience. Local food systems are less vulnerable to disruptions in global supply chains, which can be affected by several elements, including political unrest, natural disasters, or economic crises. By fostering a robust local food network, communities can ensure a more stable and reliable food supply. This resilience is critical in the face of climate change, which poses significant risks to global food production. With their focus on environmentally friendly methods, local and organic farms are more suited to adjust to shifting weather patterns and keep their communities supplied with food.

Another essential benefit of promoting organic and local agriculture is the economy. Purchasing from nearby farmers keeps more money in the neighbourhood, boosting small businesses and the local economy. This economic boom can result in the development of jobs and the revival of rural regions. Furthermore, the local farmers' propensity to reinvest their profits into the community is a significant factor in the region's general financial health. Maintaining diversified and resilient food systems depends on small farms, whose survival is aided by support for local agriculture.

Regarding health, organic and locally farmed produce frequently has more nutritional value than stuff cultivated conventionally. It is usually fresher since local produce can be collected at its ripest and quickly travelled. Fresher food keeps more of its nutrients, giving consumers more health benefits. Additionally, pesticide residues are less likely to harm human health in foods grown organically and without artificial pesticides. Research has demonstrated that some nutrients, such as antioxidants, which help prevent chronic diseases, can be found in higher concentrations in organic vegetables.

Additionally, encouraging organic and local farming strengthens the bond between food producers and customers. Through farmer's markets, farm stands, and community-supported agriculture (CSA) programs, consumers can purchase directly from farmers and gain more excellent knowledge about the people and methods used to produce their food. This link has the potential to raise awareness of and appreciation for the care and effort that goes into producing food. It also encourages people to eat with better understanding and to be more open to supporting sustainable practices.

One of the key challenges in promoting local and organic agriculture is the higher cost associated with these products. Organic farming practices, which avoid synthetic chemicals and emphasize labour-intensive methods, can result in higher production costs. Additionally, small-scale local farms do not benefit from the economies of scale that large industrial farms do, leading to higher consumer prices. However, it is essential to agriculture's actual cost of food, including conventional agriculture's environmental and social impacts. While organic and locally grown foods may have a higher upfront cost, they often offer more significant long-term benefits by preserving natural resources, supporting local economies, and promoting public health.

To increase the accessibility and affordability of these goods, policies, and programs that promote organic and local agriculture can be extremely important. Small-scale farmers can grow their operations and switch to organic practices with the support of government subsidies, farmers' and low-interest loans. Furthermore, the effectiveness and reach of local food systems can be increased by investing in local food infrastructure, such as farmers' markets, food hubs, and distribution networks. Campaigns for public awareness and educational initiatives can help enlighten customers about the advantages of organic and local farming, motivating them to choose more sustainable food choices.

Community-supported agriculture (CSA) programs support local and organic farmers while providing consumers with fresh, seasonal produce. Members of a CSA buy a portion of a farm's harvest at the beginning of the growing season and receive regular produce deliveries. This model provides farmers a stable income and helps them plan their crops more effectively. CSAs offer consumers a direct connection to the farm and a steady supply of high-quality, locally-grown food. Some CSAs include educational Farmer's, such as farm visits, cooking classes, and newsletters, that further engage members and promote sustainable food practices.

Farmers' markets are another vital component of local food systems. They provide a direct sales outlet for farmers and a convenient place for consumers to purchase fresh, locally-grown produce. Farmers' markets are also places where the community comes together to socialize, enjoy cultural events, and learn about local food production. The Supplemental Nutrition Assistance Program (SNAP) and other food assistance benefits are now accepted at many farmers' markets, increasing the accessibility of local and organic foods for low-income individuals and families.

Institutional procurement policies can also support local and organic agriculture by prioritizing purchasing these products for schools, hospitals, government agencies, and other institutions. Farm-to-school programs, for example, connect local farmers with schools to provide fresh, healthy food for students while offering educational opportunities about agriculture and nutrition. Institutional procurement can help farmers scale up their operations and achieve greater economic viability by creating a stable demand for local and organic products.

The role of consumers in supporting local and organic agriculture cannot be overstated. Consumer demand drives market trends, and individuals can influence the food system in meaningful ways by choosing to buy local and organic products. Purchasing decisions based on moral, environmental, and health factors is known as conscious consumerism, and it can encourage farmers and food producers to use more sustainable methods. Participating in community projects and advocating for legislation that supports organic and local agriculture can increase the influence of individual actions.

Growing their food is another way that customers may support sustainable agriculture in addition to buying organic and locally grown food. Home gardens, community gardens, and urban farming projects allow people to produce fruits, vegetables, and herbs. These initiatives can increase access to healthy food, particularly in urban areas and food deserts where access to fresh produce is limited. Gardening also offers educational benefits, allowing individuals to learn about plant cultivation, soil health, and sustainable practices firsthand. Community gardens, in particular, can strengthen social ties and foster a sense of community ownership and resilience.

Educational establishments play a big part in encouraging local and organic agriculture. Schools, colleges, and universities can incorporate sustainable agriculture and food systems education into their curricula, providing students with the knowledge and skills to support and advocate for these practice students. Learning opportunities like school gardens, farm internships, and service-learning projects can further enhance students' understanding and engagement. By educating the next generation about the importance of sustainable agriculture, educational institutions can help create a culture of sustainability and stewardship.

Research and innovation are also crucial for advancing local and organic agriculture. Agricultural research institutions and universities can develop and disseminate knowledge about sustainable farming practices, organic pest and disease management, soil health, and crop diversity. Innovation in areas such as organic fertilizers, natural pest control, and resilient crop varieties can improve the productivity and sustainability of local and organic farms. Agriculture's research efforts involving farmers, scientists, and policymakers can ensure that research addresses sustainable agriculture's practical needs and challenges.

In conclusion, supporting local and organic agriculture offers numerous benefits for the environment, public health, local economies, and community resilience. Organic and local farming practices help create a more equitable and sustainable food system by avoiding toxic chemicals, reducing food miles, conserving biodiversity, and enhancing soil health. Regulations, education, research, and consumer decisions are all essential to advance these practices and make them more widely available and economically feasible. People and communities can take significant steps to support local and organic farmers and establish a food system that puts fairness, sustainability, and health first as knowledge of the value of sustainable agriculture grows.

Reducing Food Waste

Food waste reduction is essential to solve issues of global food security, environmental sustainability, and economic efficiency. Food waste occurs at every stage of the food supply chain, from post-harvest handling and agricultural production to processing, distribution, retail, and customer behaviour. The amount of food lost or wasted annually is estimated to equal one-third of all food produced worldwide. This translates to over 1.3 billion tons of edibles, enough to feed billions. To address this problem, a thorough understanding of the reasons for food waste must be developed, and practical solutions must be implemented at every food chain level.

At the agricultural level, food waste can occur due to weather conditions, pests, diseases, and market fluctuations. Poor post-harvest handling and storage practices can lead to significant losses, particularly in developing countries where infrastructure and technology are often lacking. Improving agricultural practices and investing in better storage facilities can help reduce these losses. For example, using hermetically sealed storage bags can protect grains from moisture and pests, significantly extending their shelf life. Training farmers in post-harvest handling techniques and providing access to appropriate technologies can mitigate food waste at this critical stage.

Food processing and manufacturing also contribute to food waste. Inefficiencies in production processes, such as overproduction, improper calibration of machinery, and quality control issues, can lead to substantial waste. Waste can be decreased by implementing lean manufacturing concepts and making technological investments to streamline production procedures. For example, food processors can guarantee that only superior produce is processed by employing sophisticated sorting and grading technology. While this continues, inferior materials can be composted or used for animal feed. By-products from food processing, such as fruit peels and vegetable trimmings, can be repurposed into value-added products like juices, purees, and dietary supplements, thereby reducing waste and creating additional revenue streams.

The distribution and retail stages of the food supply chain are also significant sources of food waste. Transportation issues, such as delays and improper handling, can lead to spoilage, particularly for perishable goods. Improving logistics and transportation infrastructure, including better refrigeration and real-time tracking systems, can minimize these losses. At the retail level, cosmetic standards for fruits and vegetables often lead to the rejection of perfectly edible produce that needs to meet aesthetic criteria. Retailers significantly impact lowering food waste by adopting more flexible standards and educating consumers about the acceptability of "ugly" produce. Initiatives such as offering discounts on cosmetically imperfect produce and creating dedicated sections for such items can encourage consumers to purchase them, thereby reducing waste.

Consumer behaviour is one of the most significant contributors to food waste. In developed countries, much food waste occurs at the household level. Factors such as over-purchasing, improper storage, and confusion over date labelling contribute to this waste. Educating consumers about meal planning, proper food storage, and understanding date labels can significantly reduce household food waste. For instance, teaching consumers how to store fruits and vegetables correctly can extend their shelf life and prevent spoilage.

Promoting leftovers and encouraging creative cooking with surplus ingredients can also help minimize waste. Date labelling, such as "best before" and "use by" dates, often leads to confusion, with many consumers discarding food that is still safe to eat. Unnecessary waste can be decreased by making these labels clearer and encouraging the use of sensory cues (such as taste and smell) to evaluate food quality.

Food waste has profound environmental implications. When food is wasted, the resources used to produce, transport, and process that food are also wasted. This includes water, land, energy, and labour. Moreover, food waste breaks down anaerobically in landfills to generate the potent greenhouse gas methane, accelerating climate change. One of the most important methods for reducing environmental effects is to reduce food waste. Food waste can be effectively managed via composting, which converts organic waste into nutrient-rich soil amendments that improve soil health and lessen the need for synthetic fertilizers. Food waste does not need to end up in landfills by participating in household and community composting projects, which helps create a more sustainable waste management system.

Economic considerations are also significant when addressing food waste. Food waste costs the economy hundreds of billions of dollars a year. Reducing food waste can result in substantial cost savings for businesses and households. Optimizing supply chain management and implementing waste reduction strategies can improve companies' profitability. For example, food retailers can conduct regular audits to identify and address sources of waste, implement inventory management systems to minimize overstocking and donate surplus food to food banks and charities. Households can save money by planning meals more effectively, using shopping lists to avoid impulse purchases, and utilizing leftovers.

Food donation is a critical strategy for reducing food waste and addressing food insecurity. Surplus food from farms, manufacturers, retailers, and food service establishments can be redirected to those in need through food banks, soup kitchens, and other charitable organizations. Legislation such as the Good Samaritan Food Donation Act in the United States provides liability protection for food donors, encouraging more businesses to participate in food donation programs. Collaborations among corporations, nonprofit organizations, and governmental bodies can enable the equitable distribution of excess food, guaranteeing that consumable items are not thrown away and contributing to the reduction of hunger.

Novel technology and methodologies are surfacing to address food waste more efficiently. Businesses can adopt targeted interventions by identifying trends and root causes of waste with food waste tracking and analytics tools. Retail waste can be decreased by mobile apps linking customers to cheap leftover food from cafes, restaurants, and grocery shops. Furthermore, improvements in food preservation technology, like freeze-drying, vacuum sealing, and packing in a controlled atmosphere, can increase the shelf life of perishable items and decrease spoiling.

Enacting policies is essential to combating food waste. Governments may reduce waste throughout the food supply chain by enacting laws and providing incentives. For instance, setting targets for food waste reduction, mandating food waste audits for large businesses, and providing tax incentives for food donations can drive progress. Public awareness campaigns can also significantly change consumer behaviour and foster a culture of sustainability. In addition to offering helpful advice for this aim, educational initiatives in schools and community organizations can increase public awareness of the significance of minimizing food waste.

Collaborative efforts are essential for tackling food waste. Multi-stakeholder partnerships involving governments, businesses, nonprofits, and consumers can drive systemic change. Cooperative efforts, like the Champions 12.3 alliance, are a prime example. It unites leaders from many sectors to meet the UN Sustainable Development Goal of reducing global food waste per capita by 2030. Through collaboration, stakeholders can exchange optimal methodologies, generate inventive resolutions, and establish a more enduring food chain.

Cultural and societal factors also influence food waste. In some cultures, abundance and generosity are highly valued, leading to over-purchasing and over-preparation of food. Addressing food waste in these contexts requires a nuanced approach that respects cultural values while promoting more sustainable practices. Community engagement and participatory approaches can help tailor interventions to local contexts and ensure their effectiveness. For example, community-led initiatives that involve residents in food waste reduction efforts, such as gleaning programs that harvest surplus produce from farms, can build local ownership and sustainability.

In conclusion, reducing food waste is a multifaceted challenge that requires coordinated efforts across the entire food supply chain. Other methods exist to avoid problems, such as enhancing farming methods, investing in improved processing and storage technology, streamlining distribution and retail operations, and altering customer behaviour. Reducing food waste has significant advantages for the environment, the economy, and society, making it a top priority for building a more just and sustainable food system. Our ability to reduce food waste is significant and create a more sustainable future for everybody by embracing cutting-edge technologies, putting supportive laws into place, encouraging cooperation, and spreading awareness.

CHAPTER IV

Eco-Conscious Transportation

The Environmental Cost of Conventional Transportation

The environmental cost of conventional transportation is a multifaceted issue with significant implications for the planet's health and sustainability. Conventional transportation primarily uses internal combustion engine (ICE) cars powered by fossil fuels like gasoline, diesel, and natural gas. This mode of transportation dominates global mobility, including cars, trucks, buses, airplanes, and ships. While conventional transportation has facilitated economic growth, connectivity, and convenience, it also poses severe environmental challenges. These challenges include air pollution, greenhouse gas emissions, resource depletion, habitat destruction, and the broader impacts on public health and ecosystems.

Air pollution is one of the most pressing environmental costs associated with conventional transportation. ICE vehicles emit a variety of harmful pollutants, including nitrogen oxides (NOx), carbon monoxide (CO), volatile organic compounds (VOCs), particulate matter (PM), and sulphur dioxide (SO_2). Due to their role in creating smog, acid rain, and ground-level ozone, these pollutants harm the environment and human health. For example, sunlight causes nitrogen oxides and volatile organic compounds to react, forming ozone, a major component of smog that aggravates asthma and can cause cardiovascular and respiratory issues. Heart attacks, strokes, and lung cancer are just a few of the health problems that can be brought on by particulate matter or excellent particles (PM2.5), which can enter the bloodstream and enter the lungs deeply. Moreover, sulphur dioxide can cause acid rain, which damages flora, soils, and aquatic habitats.

Greenhouse gas emissions from conventional transportation are a significant driver of climate change. Transport is one of the main sectors that emit CO2 into the atmosphere, accounting for about 24 percent of global emissions from the combustion of fossil fuels. Significant volumes of CO2 are released while burning gasoline and diesel fuels, which build up in the atmosphere and intensify the greenhouse effect, causing global warming. In addition to CO2, other greenhouse gases released by transportation include nitrous oxide (N2O) and methane (CH4), which have a higher potential to contribute to global warming. As a result of these emissions, there has been an increase in the frequency and power of extreme weather events, as well as a rise in sea level and global temperatures. These climatic changes to human and ecological systems, such as infrastructure, agriculture, water supplies, and health, pose serious hazards.

Resource depletion is another significant environmental cost of conventional transportation. Obtaining, refining, and using fossil fuels for transportation deplete finite natural resources and cause ecological damage. Oil extraction, for example, can lead to habitat destruction, soil and water contamination, and biodiversity loss. Due to their environmental impacts, offshore drilling and hydraulic fracturing (fracking) are particularly controversial oil extraction methods. Offshore drilling can result in oil spills, which devastate marine ecosystems, killing wildlife and damaging habitats. Fracking, conversely, involves injecting high-pressure fluid into underground rock formations to release oil and gas, which can contaminate groundwater, induce earthquakes, and release methane into the atmosphere.

The production and disposal of vehicles also contribute to environmental degradation. Manufacturing cars involves extracting and processing raw materials such as steel, aluminium, and plastics, which need a large quantity of water and energy. Mining activities associated with these materials can lead to deforestation, habitat destruction, and soil and water pollution. Moreover, the disposal of vehicles at the end of their life cycle generates substantial waste, including hazardous materials such as lead-acid batteries, motor oil, and coolant. To reduce environmental harm, these items must be recycled and disposed of properly. Nevertheless, automobiles often wind up in landfills, where they can release harmful materials into the groundwater and soil.

Conventional transportation also contributes to habitat destruction and fragmentation. The construction of roads, highways, and other transportation infrastructure often requires clearing forests, wetlands, and other natural habitats. For wildlife, this fragmentation and loss of habitat can have dire repercussions, including dwindling species numbers, declining biodiversity, and disruption of ecosystem services. Highways and roads can impede the movement of animals, making it more difficult for them to find food, partners, and nesting locations. They can also raise the possibility of car accidents. Infrastructure related to transportation can also change hydrological patterns, raising the possibility of flooding and soil erosion.

The environmental cost of conventional transportation extends to the aquatic environment through water pollution. Road salts, oil, grease, heavy metals, and other contaminants can enter adjacent water bodies through runoff from roads and highways. This discharge can destroy aquatic life and ecosystems and lower water quality. Oil spills from ships and offshore drilling operations seriously threaten marine environments, which harm marine organisms and habitats over time. Another worry is the introduction of alien species by ship ballast water discharge. This may cause native species to be displaced and marine ecosystems to be disrupted.

Noisy traffic has a significant but less apparent negative impact on the environment. The noise made by cars, trains, and airplanes can have a detrimental effect on both humans and animals. People who regularly expose themselves to high levels of vehicle noise may suffer from stress-related ailments, sleep disturbances, cardiovascular problems, and hearing loss. Among the ways noise pollution affects wildlife are disruptions to mating calls, hunting, communication, and predator avoidance. Marine mammals, like dolphins and whales, depend on echolocation for hunting and navigation; however, noise from ships and undersea drilling operations can disturb these vital activities.

The environmental costs of conventional transportation significantly impact public health. Air pollution from vehicle emissions is a significant health risk, contributing to respiratory and cardiovascular diseases, premature deaths, and reduced quality of life. Children, the elderly, and people with pre-existing medical disorders are among the vulnerable groups that are most vulnerable. Pollution concentrations are high in metropolitan areas, where traffic congestion is predicted, exacerbating health issues and raising healthcare expenditures.

The environmental costs of traditional transportation must be addressed through a multimodal strategy incorporating policy changes, technological innovation, behavioural adjustments, and infrastructure upgrades. Making the switch to greener, more sustainable forms of transportation is crucial to reducing the adverse effects on the environment. Since they can run on renewable energy sources and emit no tailpipe emissions, electric cars (EVs) and hydrogen fuel cell vehicles are attractive substitutes for traditional internal combustion engine (ICE) vehicles. Nevertheless, the energy mix utilized to produce hydrogen and power determines how beneficial these technologies are for the environment. To fully maximize the environmental benefits of hydrogen fuel cell vehicles and electric vehicles (EVs), it is imperative to ensure that the electrical grid receives increasing power from hydroelectric electricity, solar energy, and wind energy, among other green sources.

Policies are essential for advancing environmentally friendly transportation. Governments can impose rules and incentives, such as tax breaks for low-emission car sales, subsidies for EV purchases, and stricter emissions requirements for conventional automobiles, to promote the adoption of greener vehicles. Investing in public transportation infrastructure is another essential tactic to lessen dependency on private cars and cut emissions overall. Public transportation networks, including buses, trains, and subways, can be expanded and improved to offer convenient and reasonably priced driving alternatives, lowering pollution and traffic jams. Benefits to the environment and public health can also be significantly increased by policies that encourage cycling and walking as active modes of transportation.

Infrastructure that is safe and accessible can encourage more people to use these eco-friendly modes of transportation, including bike lanes, pedestrian walkways, and bike share programs.

Urban planning and land-use policies are also crucial in addressing the environmental costs of transportation. Designing cities and communities to be more compact and mixed-use can reduce the need for long commutes and encourage public transit, walking, and cycling. Transit-oriented development, which focuses on creating dense, mixed-use communities around public transit hubs, can reduce car dependency and promote sustainable mobility. Implementing zoning regulations that support higher density and mixed land use can create more liveable and sustainable urban environments.

Technological innovations and intelligent mobility solutions can also help reduce transportation's environmental impact. The development of autonomous vehicles, for example, has the potential to improve traffic efficiency, reduce accidents, and lower emissions. Intelligent and connected transportation systems use real-time data and communication technologies to optimize traffic flow, lessen congestion, and consume less fuel. Mobile apps-enabled car and ride-sharing services can lower the number of vehicles on the road and encourage more economical use of resources.

Public awareness and behavioural change are essential components of reducing transportation's environmental costs. Educating the public about conventional transportation's ecological and health impacts and promoting sustainable transportation options can drive behavioural shifts. Campaigns that encourage carpooling, the use of public transit, and the adoption of cleaner vehicles can influence individual choices and contribute to broader societal change. Additionally, promoting telecommuting and flexible work arrangements can reduce the need for daily commuting and lower overall transportation emissions.

International coordination and collaboration are also essential to address the global character of transportation emissions. Air pollution and climate change are transboundary problems that call for cooperation across nations and regions. Setti's goals and standards for lowering transportation-related emissions greatly aid organizations ' internal frameworks and agreements like the Paris Agreement and the International Maritime Organization's shipping emissions regulations. Nations exchanging best practices, technology, and regulations can speed up the shift to more environmentally friendly transportation systems.

In conclusion, the environmental cost of conventional transportation is a significant and multifaceted challenge that encompasses air pollution, greenhouse gas emissions, resource depletion, habitat destruction, water pollution, noise pollution, and public health impacts. Addressing these challenges requires a comprehensive and integrated approach that includes technological innovation, policy measures, urban planning, public awareness, and international cooperation. Transitioning to cleaner and more sustainable transportation modes, improving public transit and active transportation infrastructure, and promoting behavioural changes are essential steps in mitigating the environmental impacts of transportation. By adopting these strategies, we can create a more sustainable and resilient transportation system that supports ecological health, economic prosperity, and social well-being.

Green Transportation Options

Green transportation options are crucial for creating a sustainable and environmentally friendly future. As concerns about climate change, air pollution, and resource depletion grow, transitioning from conventional transportation methods, which heavily rely on fossil fuels, to greener alternatives becomes more pressing. Green transportation encompasses many strategies and technologies to reduce the environmental impact of moving people and goods. These include electric vehicles (EVs), hydrogen fuel cell vehicles, public transportation, cycling, walking, car-sharing, and advancements in fuel efficiency and alternative fuels. Implementing and promoting these options can significantly decrease greenhouse gas emissions, improve air quality, conserve natural resources, and enhance public health.

EVs are the cutting edge of environmentally friendly transportation. EVs have no tailpipe emissions and are powered by electricity stored in batteries instead of internal combustion engine (ICE) cars. This makes them an effective tool for reducing air pollution and greenhouse gas emissions, mainly when the electricity required to charge them is produced through clean energy sources like hydroelectricity, wind, or solar energy. Technological developments in batteries have sped up the adoption of electric vehicles (EVs) by increasing their range, reducing charging times, and lowering prices. Incentives, including tax credits, rebates, and grants, are being offered by governments worldwide to promote the purchase of electric vehicles. Additionally, EVs are now more convenient owing to the expansion of the charging infrastructure, which includes fast-charging networks accessible to a more excellent range of users.

Hydrogen fuel cell vehicles represent another promising green transportation option. These cars use hydrogen gas in a fuel cell to react chemically with oxygen to create energy; the only byproduct they release is water vapor. Like conventional gasoline vehicles, hydrogen fuel cell vehicles have the advantage of quick refilling periods and longer driving ranges. Moreover, hydrogen can be produced from various sources, including renewable energy through electrolysis, which splits water into hydrogen and oxygen. However, obstacles like the high cost of fuel cells and the requirement for a comprehensive infrastructure for hydrogen refuelling stand in the way of the extensive use of hydrogen fuel cell automobiles. Notwithstanding these obstacles, continuous study and development are bringing hydrogen closer to being a practical and sustainable replacement for fossil fuels.

Public transit is essential for minimizing the environmental impact of urban travel. Large groups of people can be moved effectively by buses, trains, trams, and subways, which lowers the number of individual cars on the road and, consequently, overall emissions. Electrification of public transit systems, such as electric buses and trains, further enhances their environmental benefits by minimizing reliance on fossil fuels. Cities worldwide are investing in expanding and modernizing their public transportation networks to make them more reliable, convenient, and attractive to commuters. Integrating public transportation with other modes of green transportation, such as bike-sharing and walking, can create seamless and sustainable urban mobility solutions. Additionally, policies such as congestion pricing and dedicated bus lanes can improve the efficiency and appeal of public transit, encouraging more people to opt for these greener options.

Walking and riding are the most eco-friendly forms of transportation transportation, as they produce no emissions and have numerous health benefits. Promoting active transportation requires creating safe and accessible infrastructure, including bike lanes, pedestrian paths, and bike-sharing programs. Cities like Copenhagen and Amsterdam have successfully implemented extensive cycling infrastructure, making cycling a primary mode of transport for many residents. These cities provide valuable models for how urban planning and policy can prioritize active transportation. Encouraging cycling and walking can reduce traffic congestion, lower transportation costs, and improve public health by promoting physical activity. Moreover, integrating cycling and walking with public transportation can enhance urban mobility systems' overall sustainability and efficiency.

Services like ride- and car-sharing are creative ways to reduce the number of cars on the road and encourage more effective use of available resources. Car-sharing programs, such as Zipcar and Car2Go, allow members to rent vehicles in the short term in short term, reducing the need for car ownership and encouraging alternative transportation modes for daily commutes. Ride-sharing platforms like Uber and Lyft match passengers with drivers heading in the same direction, maximizing vehicle occupancy and reducing the number of single-occupancy trips. These services can complement public transportation and provide flexible, on-demand mobility options. Additionally, integrating electric and hybrid vehicles into car-sharing and ride-sharing fleets can further reduce their environmental impact.

Advancements in fuel efficiency and alternative fuels are also essential for green transportation. Improving the fuel economy of conventional vehicles through technological innovations, such as turbocharging, direct fuel injection, and lightweight materials, can significantly reduce fuel consumption and emissions. Hybrid cars, which integrate an electric motor and internal combustion engine, offer improved fuel efficiency and lower emissions than traditional gasoline vehicles. Additionally, alternative fuels such as biofuels derived from renewable biological resources like plants and algae can provide a sustainable and lower-carbon alternative to fossil fuels. The development and adoption of advanced biofuels, such as cellulosic ethanol and algae-based biodiesel, hold promise for further reducing the carbon footprint of transportation.

Urban planning and land-use policies are vital for supporting green transportation. Designing cities and communities to be more compact and mixed-use can reduce the need for long commutes and make walking, cycling, and public transportation more viable and attractive options. Transit-oriented development (TOD), which focuses on creating dense, mixed-use communities around public transit hubs, can reduce car dependency and promote sustainable urban mobility. Implementing zoning regulations that support higher density and mixed land use can create more liveable and sustainable urban environments. Green areas and pedestrian-friendly layouts can also improve the standard of living and promote the use of more environmentally friendly forms of transportation.

Government incentives and laws greatly aid green transportation. A variety of policies can be put in place by policymakers to promote the use of more environmentally friendly transportation options and discourage the use of fossil fuel-powered cars. These policies include tax breaks and subsidies for electric and hybrid vehicles, funding for infrastructure construction for fuelling and charging, and rules imposing strict limits on vehicle emissions. Governments can also fund R&D to advance new technology and facilitate pilot projects showing green transportation solutions' viability and advantages. Public awareness campaigns and educational initiatives can effectively encourage people and businesses to adopt greener decisions and promote sustainable mobility practices.

The role of the private sector in advancing green transportation must be considered. Automakers, technology companies, and transportation providers are crucial players in developing and deploying innovative solutions that reduce the environmental impact of transportation. For instance, significant automakers invest heavily in developing electric and hydrogen fuel cell vehicles, while technology companies are creating intelligent mobility solutions that optimize transportation systems and reduce emissions. Collaboration between the public and private sectors is essential to accelerate the transition to green transportation and ensure that sustainable mobility solutions are widely adopted and accessible.

International cooperation and coordination are also vital for addressing the global nature of transportation emissions. Climate change and air pollution are transboundary issues that require collaborative efforts across countries and regions. International institutions and accords, like the Paris Agreement and the International Maritime Organization's regulations on shipping emissions, play a crucial role in setting targets and standards for reducing transportation-related emissions. Sharing best practices, technologies, and policies among countries can accelerate the transition to more sustainable transportation systems. Additionally, international financial institutions and development agencies can support developing countries in adopting green transportation solutions by providing funding, technical assistance, and capacity-building programs.

Public awareness and education are essential components of promoting green transportation. Educating the public about conventional transportation's environmental and health impacts and the benefits of greener alternatives can drive behavioural changes and increase demand for sustainable transportation options. Campaigns that encourage carpooling, the use of public transit, and the adoption of cleaner vehicles can influence individual choices and contribute to broader societal change. Additionally, promoting telecommuting and flexible work arrangements can reduce the need for daily commuting and lower overall transportation emissions. Educational institutions, universities, and community organizations can significantly promote sustainability and awareness-raising by including lessons on green mobility in their curricula and activities.

The transition to green transportation offers numerous benefits beyond environmental sustainability. It can create economic opportunities by driving innovation and growth in the clean technology and renewable energy sectors. By lowering reliance on imported fossil fuels, investments in green transportation infrastructure and technologies can improve economic growth, create jobs, and increase energy security. Furthermore, reducing emissions and enhancing air quality can substantially impact community well-being, public health, and healthcare expenditures. We can build a future for everyone that is more robust, sustainable, and egalitarian by adopting green transportation.

In conclusion, green transportation options are essential for addressing the environmental challenges posed by conventional transportation. Electric vehicles, hydrogen fuel cell vehicles, public transit, cycling, walking, car-sharing, and advancements in fuel efficiency and alternative fuels all offer viable pathways to reducing the environmental impact of moving people and goods. Implementing these options requires a comprehensive and integrated approach that includes technological innovation, policy measures, urban planning, public awareness, and international cooperation. By transitioning to greener transportation modes, we can significantly decrease greenhouse gas emissions, improve air quality, conserve natural resources, and enhance public health. Building a sustainable and ecologically friendly transportation system that promotes the welfare of present and future generations will require the combined efforts of governments, corporations, communities, and individuals.

Car-Free Living

Car-free living is a lifestyle choice that seeks to eliminate the dependence on personal automobiles for transportation, opting for alternative means such as walking, cycling, public transit, and car-sharing. This lifestyle addresses environmental concerns and promotes a healthier, more sustainable, socially connected lifestyle. The concept of car-free living can be traced back to the early 20th century. Still, it has gained significant traction in recent years due to increased awareness of urban overcrowding, climate change, and the need for a higher standard of living.

Living without a car can dramatically lower emissions of particulate matter (PM), carbon dioxide (CO_2), nitrogen oxides (NOx), and other pollutants by reducing the number of cars on the road. By lowering emissions, we can slow down climate change, enhance air quality, and lower the prevalence of heart and lung conditions caused by air pollution. Less reliance on automobiles also reduces the need for fossil fuels, which lessens the harm that transportation, refining, and oil extraction due to the environment.

Urban areas often suffer from severe traffic congestion, leading to wasted time, increased stress, and economic losses. Car-free living can alleviate these problems by reducing the number of vehicles on the road. Fewer cars mean less traffic congestion, shorter travel times, and more efficient movement of people and goods. Moreover, reduced traffic can lead to fewer accidents, lower road maintenance costs, and decreased noise pollution, all contributing to a more pleasant urban environment. Cities designed with car-free principles often prioritize pedestrian-friendly infrastructure, creating vibrant, accessible public spaces that encourage social interaction and community engagement.

Public health is another significant area where car-free living can positively impact. Walking and cycling are great exercises that can help avoid heart disease, diabetes, obesity, and other lifestyle-related illnesses. Encouraging these modes of transportation can lead to a more active population, reducing healthcare costs and improving overall well-being. Moreover, reduced air pollution from fewer cars can decrease the prevalence of asthma, lung cancer, and other respiratory conditions, particularly in children and the elderly, who are more vulnerable to poor air quality.

Social benefits of car-free living include increased social cohesion and a stronger sense of community. Car-free areas often feature more public spaces where people can gather, interact, and participate in community activities. These areas can promote community and enhance mental health by reducing isolation and encouraging social interaction. Additionally, car-free living can make cities more inclusive by improving accessibility for people with disabilities, older people, and children, who may find car-centric environments challenging to navigate.

Living without a car has significant financial benefits. Families without automobiles can save much money on parking, insurance, gasoline, and maintenance. Local economies can be strengthened by directing these savings toward other necessities or discretionary consumption. Cities that invest in public transit and car-free infrastructure can achieve incredible economic growth and development, as these factors draw in businesses, visitors, and people looking for a high standard of living.

Implementing car-free living requires thoughtful urban planning and policy measures. Cities must develop and maintain comprehensive public transportation systems that are reliable, affordable, and accessible. Investments in infrastructure such as bike lanes, pedestrian paths, and public transit networks are crucial to making car-free living a viable option. Moreover, policies such as congestion pricing, car-free zones, and incentives for using alternative transportation can encourage residents to shift away from car dependency.

When living without a car, make sure there are convenient and readily available alternatives to driving. Public transportation systems must be extensive and efficient, covering a wide area and offering frequent, timely service. Cities like Copenhagen, Amsterdam, and Zurich have successfully implemented such systems, becoming models for car-free living. These cities provide robust networks of buses, trams, and trains, complemented by extensive cycling infrastructure and pedestrian-friendly streets.

Cycling infrastructure plays a critical role in supporting car-free living. Bike-sharing schemes, designated bike lanes, and safe places to park can help more people decide to ride their bikes instead of driving cars. Cities with well-developed cycling infrastructure, such as Amsterdam and Copenhagen, have high rates of cycling and lower levels of car dependency. These cities also invest in safety measures, such as traffic calming and bike-friendly traffic signals, to make cycling safer and more attractive.

Pedestrian-friendly design is another essential component of car-free living. This involves creating walkable neighbourhoods with safe, accessible sidewalks, pedestrian crossings, and public spaces. Walking can be more convenient and enjoyable when residential, commercial, and recreational amenities are integrated into a mixed-use complex. This reduces the need for long trips. By prioritizing pedestrian infrastructure, cities can develop thriving, liveable neighbourhoods where people can get most of their everyday necessities without a car.

Car-sharing and ride-sharing services can also support car-free living by providing flexible transportation options when a car is necessary. Car-sharing programs, such as Zipcar and Car2Go, allow members to rent vehicles for short periods, reducing the need for car ownership. Ride-sharing platforms like Uber and Lyft can provide convenient, on-demand transportation, particularly in areas with limited public transit options. These services can complement public transportation and active transportation modes, making it easier for people to live without owning a car.

Working remotely and telecommuting has grown in popularity, especially after the COVID-19 epidemic. These work arrangements can reduce the need for daily commuting and support car-free living by allowing people to work from home or in local coworking spaces. Companies that offer flexible work options can contribute to reducing traffic congestion, lowering emissions, and improving work-life balance for their employees.

Educational campaigns and community engagement are essential for promoting car-free living and encouraging people to adopt sustainable transport habits. Public awareness campaigns can highlight the environmental, health, and economic benefits of car-free living, while community programs can provide support and resources for those making the transition. Educational institutions, colleges, and community-based groups can contribute significantly to public education and sustainable culture development.

Government policies and regulations are crucial for supporting car-free living. Policymakers can implement various measures to encourage alternative transportation and discourage car dependency. These measures include investments in public transit infrastructure, subsidies for electric bikes and scooters, and incentives for businesses to promote sustainable commuting options. Additionally, governments can introduce regulations such as low-emission zones, congestion pricing, and parking restrictions to reduce the number of cars on the road.

The concept of car-free cities is gaining traction worldwide, with several cities implementing car-free zones and car-free days to promote sustainable urban mobility. For example, Oslo has introduced car-free zones in its city centre, prioritizing pedestrians, cyclists, and public transit. Similar to this, Paris has increased the number of bike lanes and instituted car-free days, which have improved accessibility and enjoyment for both locals and tourists. These programs show how living without a car may enhance urban settings and people's quality of life.

In addition to environmental and social benefits, car-free living can enhance urban resilience. By relying more on cars, cities can adapt to changing circumstances, such as rising fuel prices, economic fluctuations, and natural disasters. Car-fire infrastructure, such as pedestrian paths and bike lanes, can provide alternative transportation routes during emergencies, improving the city's ability to respond to and recover from crises.

Living without a car also lessens the impact of the urban heat island effect, which is the phenomenon wherein cities get hotter than rural regions because of the concentration of infrastructure, roads, and other structures. Vehicles add to this effect by releasing pollution nd heat. Reducing the number of automobiles may increase green space, lower temperatures, and generate better air quality and living conditions.

The success of car-free living depends on a collaborative effort involving governments, businesses, communities, and individuals. Governments must provide the necessary infrastructure and policy support, while companies can promote sustainable commuting options and invest in green technologies. Communities can advocate for car-free initiatives and support local projects that enhance walkability and bike ability. People can consciously try to use fewer cars and, whenever possible, choose environmentally friendly modes of transportation.

Innovations in technology are equally vital to the cause of car-free life. The use of alternate forms of transportation is growing in popularity and practicality because of innovations like intelligent transportation systems, driverless cars, and electric bikes and scooters. Electric bikes and scooters are helpful for short trips, especially in steep terrain, or for individuals who might find regular cycling difficult. Self-driving cars have the potential to transform car-sharing and public transit, improving its usability and efficiency. Intelligent transportation systems, which use data and technology to optimize traffic flow and public transit schedules, can improve the efficiency and reliability of sustainable transportation options.

The transition to car-free living can be challenging, particularly in regions where car dependency is deeply ingrained in the culture and infrastructure. However, the benefit of this lifestyle choice makes it a worthwhile endeavour. Reducing our reliance on cars can create more sustainable, healthy, and vibrant communities that are better equipped to face future challenges. Embracing car-free living requires a shift in mindset and a willingness to explore and invest in alternative transportation options. With the right policies, infrastructure, and community support, car-free living can become a viable and attractive option for people worldwide.

Cities must be designed with accessibility and sustainability in mind, prioritizing mixed-use development, high-density housing, and public transportation. Compact, walkable neighbourhoods can reduce the need for long commutes and make alternative transportation options more practical and appealing. Additionally, integrating green spaces, parks, and pedestrian-friendly streets into urban design can enhance the quality of life and promote social interaction.

The economic benefits of car-free living extend beyond individual savings on car-related expenses. On a broader scale, cities that invest in sustainable transportation infrastructure can attract businesses, residents, and tourists seeking a high quality of life. These cities often experience economic growth and development due to their commitment to sustainability. Furthermore, reducing traffic congestion and improving air quality can enhance a city's overall attractiveness, making it a desirable place to live and work.

Car-free living can also contribute to social equity by improving access to transportation for all residents. In car-dependent cities, those who cannot afford a car or are unable to drive may face significant barriers to accessing employment, education, healthcare, and other essential services. Cities may guarantee that all citizens can access dependable, reasonably priced, and practical transportation options by investing in public transportation and infrastructure for active transportation. This can support inclusive, equitable societies and lessen social inequality.

The role of technology in supporting car-free living cannot be overstated. Innovations such as mobility-as-a-service (Maas) platforms, which integrate various transportation modes into a user-friendly app, can make planning and pay for their trips easier by using public transit, bike-sharing, and car-sharing services. Real-time data and smart city technologies can optimize traffic flow, reduce wait times, and improve the overall efficiency of transportation systems. Additionally, the development of electric and autonomous vehicles can enhance the sustainability and convenience of alternative transportation options.

The promotion of car-free living must include both education and awareness-raising. Campaigns for public awareness can draw attention to the advantages of environmentally friendly transportation and inspire individuals to live car-free lives. Schools and universities can integrate sustainability topics into their curricula, teaching students about transportation choices' environmental, mental, and social impacts. Community organizations can organize events, workshops, and initiatives promoting car-free living and providing resources and support for transitioning.

The concept of car-free living is not new, but it has gained renewed interest and momentum in recent years. Cities worldwide recognize the need to reduce car dependency and are implementing innovative solutions to promote sustainable transportation. For example, Freiburg in Germany has pioneered car-free living, with its Vauban district designed to prioritize pedestrians, cy lists, and public transit. Similarly, Pontevedra, Spain, increased the quality of life for its citizens by making its city centre a car-free zone and significantly improving the air and noise pollution.

The COVID-19 pandemic has also significantly impacted transportation patterns, highlighting the potential for car-free living. With lockdowns and social distancing measures in place, many people have turned to walking and cycling for their daily exercise and essential trips. Public transportation usage declined, but there was a notable increase in the use of bikes and scooters. Cities responded by creating temporary bike lanes and pedestrian zones, demonstrating the feasibility and benefits of car-free living. As we move forward, we can build on these changes and create more permanent solutions that support sustainable transportation.

The future of car-free living depends on our ability to create resilient, adaptable, and sustainable urban environments. This requires a holistic approach integrating transportation planning with land use, housing, economic development, and environmental protection. By prioritizing sustainable mobility options, making infrastructure investments, and cultivating a culture of sustainability, we can build more liveable, fair, dynamic, and ecologically friendly communities.

In conclusion, car-free living offers numerous benefits for the environment, public health, social cohesion, and the economy. We can create more sustainable, healthy, and inclusive communities by reducing our reliance on personal automobiles and embracing alternative transportation options. Achieving car-free living requires a comprehensive approach that includes urban planning, policy measures, technological innovations, and public awareness. Car-free living can become a feasible and appealing alternative for people worldwide with the proper support and dedication, assisting in addressing some of the most critical issues of our day.

Tips for Reducing Transportation Footprint

Reducing your transportation footprint is essential for a more sustainable and environmentally friendly lifestyle. Considering that personal automobiles account for the majority of transportation-related greenhouse gas emissions, improving this area of our daily lives can substantially impact air quality, climate change, and general health. Individuals and communities can adopt numerous strategies and habits to minimize their transportation-related environmental impact. These range from choosing alternative modes of transport, improving vehicle efficiency, and adopting digital solutions to making lifestyle changes that reduce the need for travel.

One of the most effective ways to reduce your transportation footprint is to embrace active transportation modes such as walking and cycling. These techniques have no emissions and provide various health advantages, such as better mental, physical, and cardiovascular health. Creating a routine that includes walking or cycling for short trips can drastically reduce car usage. For instance, commuting to work by bike or on foot lowers carbon emissions and alleviates traffic congestion and noise pollution. Additionally, urban areas can support this shift by investing in pedestrian-friendly infrastructure, such as safe sidewalks, bike lanes, and secure bike parking facilities. Cities that prioritize these infrastructures, like Copenhagen and Amsterdam, serve as models by showing how a significant proportion of the population can rely on cycling for daily commutes.

Public transportation is another critical component in reducing the transportation footprint. Buses, trains, trams, and subways can move large numbers of people efficiently and with lower per capita emissions than personal vehicles. Utilizing public transport for daily commutes can significantly cut down individual carbon footprints. To make public transportation more appealing, municipalities need to invest in reliable, frequent, and accessible services. Enhancing the convenience and comfort of public transport can shift public perception and usage patterns. For example, integrated ticketing systems, real-time updates, and well-maintained transit stations can improve user experience and encourage more people to opt for public transit over driving.

Carpooling and ride-sharing are practical solutions for those who cannot avoid using a car. By sharing rides with others, individuals can reduce the number of vehicles on the road, leading to lower overall emissions and less traffic congestion. Services like Uber Pool and Lyft Line and traditional carpooling arrangements among colleagues and neighbours can make this option convenient and cost-effective. Employers can support carpooling by providing incentives such as preferred parking spots for carpool vehicles or organizing company-sponsored carpool programs. Technology platforms that match drivers with passengers heading in the same direction can facilitate efficient ride-sharing arrangements, reducing the need for single-occupancy vehicle trips.

Choosing a fuel-efficient or electric vehicle (EV) can significantly reduce the environmental impact for those who need to drive. EVs produce zero tailpipe emissions and can offer a truly green alternative to gasoline-powered cars when charged with renewable energy. The market for EVs is rapidly expanding, with improvements in battery technology increasing range and reducing charging times. Governments can promote electric vehicles (EVs) by providing grants, tax credits, and rebates and funding the infrastructure required to charge EVs. Additionally, hybrid cars, which combine an internal combustion engine with an electric motor, provide a transitional option for those not yet ready to switch entirely to electric. These vehicles offer better fuel efficiency and lower emissions than traditional gasoline cars.

Improving driving habits can also play a significant role in reducing the transportation footprint. Eco-driving techniques, such as maintaining steady speeds, avoiding rapid acceleration and braking, and keeping tires properly inflated, can enhance fuel efficiency and reduce emissions. Regular vehicle maintenance, including oil changes and air filter replacements, ensures a car runs optimally and produces fewer pollutants. Drivers can also minimize idling, as idling consumes fuel and generates unnecessary emissions. For instance, if you stop the engine after over a minute, you can conserve gasoline and lessen your environmental influence. Eco-friendly driving behaviours can become commonplace if drivers know about these activities through public awareness campaigns and training programs.

Telecommuting and remote work have emerged as powerful tools in reducing transportation-related emissions. The CO-19 pandemic demonstrated that many jobs can be performed effectively from home, reducing the need for daily commutes and thus lowering transportation footprints. Employers can support telecommuting by adopting flexible work policies, providing necessary technological tools, and fostering a culture that values productivity over physical presence. Even a hybrid model, where employees work from home a few days a week, can significantly reduce commuting-related emissions. Additionally, virtual meetings and online collaboration tools can reduce the need for business travel, reducing transportation impacts.

Opting for sustainable travel choices can make a significant difference for those who travel frequently. Public transit, like buses and trains, is usually more environmentally friendly when traveling between cities than driving or flying. When air travel is unavoidable, choosing direct flights, which have lower emissions per mile than connecting flights, can reduce the environment's impact. Some airlines also offer carbon offset programs, allowing passengers to compensate for their flight emissions by investing in environmental projects. Additionally, promoting and using alternative travel methods, such as high-speed rail, which is more efficient and less polluting than short-haul flights, can help mitigate the environmental impact of long-distance travel.

Reducing transportation footprints also involves rethinking how goods are transported. Supporting local and regional products can reduce the emissions associated with long-distance shipping. Locally sourced goods do not have to travel as far, which cuts down on the fuel used and the pollutants emitted. Farmers' markets, local businesses, and community-supported agriculture (CSA) programs can provide consumers with fresh, locally produced goods, reducing the need for extensive transportation networks. Additionally, companies can optimize their supply chains to minimize transportation distances and use more sustainable modes of transport, such as rail or electric delivery vehicles, to reduce their environmental impact.

Urban planning and design play crucial roles in facilitating reduced transportation footprints. Creating mixed-use developments with residential, commercial, and recreational facilities within walking or cycling distance can reduce the need for long commutes. Transit-oriented development (TOD), which focuses on high-density development around public transit hubs, can make it easier for residents to rely on public transportation instead of personal vehicles. Incorporating green spaces, pedestrian zones, and cycling paths into urban planning can create more liveable, sustainable cities where residents can easily access essential services and amenities without driving.

Community initiatives can also support efforts to reduce transportation footprints. Car-free zones, pedestrian streets, and bike-sharing programs can encourage residents to adopt more sustainable transportation habits. Community events, such as car-free days and bike-to-work weeks, can raise awareness about the benefits of reducing car usage and promote alternative transportation options. Local governments and organizations can collaborate to develop and implement policies and programs supporting sustainable transportation, such as subsidies for public transit passes, bike-sharing programs, and incentives for businesses to promote sustainable commuting options.

Investing in renewable energy infrastructure is another critical strategy for reducing transportation footprints. By using renewable energy sources like hydroelectric, solar, and wind power to power electric cars and public transit networks, we may further lessen transportation's environmental impact. Governments and utilities can support this transition by investing in renewable energy projects, expanding the grid capacity, and providing incentives for adopting renewable energy technologies. Individuals and businesses can also choose renewable energy options for their homes and offices, reducing their overall carbon footprint.

Encouraging the widespread adoption of sustainable transportation methods requires the implementation of educational initiatives and public awareness campaigns. Students and locals can learn about the advantages of alternative transportation options and the effects of transportation on the environment from schools, universities, and community organizations. Public awareness campaigns can emphasize the advanced age of reducing car use and embracing sustainable transportation practices for the environment, economy, and health. These initiatives can empower people and communities to make more environmentally responsible transportation decisions by fostering a sustainable culture and providing the necessary knowledge and tools.

Implementing technology and innovation in transportation systems can also help reduce environmental impacts. Intelligent transportation systems, which use data and technology to optimize traffic flow and public transportation schedules, can increase the efficiency and dependability of transportation networks. Technology advancements like self-driving cars, electric scooters, and bike-sharing schemes can offer handy and environmentally friendly transportation choices. Furthermore, improvements in battery technology and the incorporation of renewable energy sources can improve the sustainability of electric vehicles and public transportation networks.

Supporting initiatives to lessen transportation footprints requires laws and regulations. Governments can implement a variety of policies, including regulating vehicle emissions, offering financial incentives for the use of public transit and electric cars, and making investments in environmentally friendly infrastructure.

Parking restrictions, low-emission zones, and congestion pricing can all discourage driving and promote other modes of transportation. Governments may spearhead the shift to more environmentally friendly transportation systems by establishing a conducive policy environment.

International cooperation and coordination are also essential for addressing the global nature of transportation emissions. Climate change and air pollution are transboundary issues that require collaborative efforts across countries and regions. Setting goals and standards for cutting emissions related to transportation is made possible by international accords and frameworks like the Paris Agreement and the shipping emissions regulations of the International Maritime Organization. Sharing best practices, technologies, and policies among countries can accelerate the transition to more sustainable transportation systems. Additionally, international financial institutions and development agencies can support developing countries in adopting green trans rotation solutions by providing funding, technical assistance, and capacity-building programs.

Additionally, the private sector is a major player in reducing transportation footprints. Businesses can adopt sustainable transportation policies, such as promoting telecommuting, providing incentives for employees to use public transportation, and investing in electric vehicle fleets. Companies involved in logistics and delivery can optimize their supply chains, use more efficient modes of transport, and invest in green technologies.

Collaboration between the public and private sectors is essential to drive innovation, investment, and implementation of sustainable transportation solutions.

Addressing transportation footprints also involves rethinking consumer behaviour and lifestyle choices. People can reduce their use of cars by walking or cycling short distances, consolidating errands into a single trip, and taking advantage of car-sharing or ride-sharing services. Customers can also support companies and goods that prioritize sustainability, such as those that employ locally sourced and sustainably derived materials.

CHAPTER V

Sustainable Fashion

The Environmental Impact of the Fashion Industry

Under the glitter of runway displays and glossy magazine covers, the fashion industry has a tremendous environmental footprint that frequently goes unnoticed despite its glamour and attraction. Every stage of the fashion supply chain impacts the environment, from extracting raw materials to manufacturing, shipping, and disposing of clothing. The most important factor contributing to environmental degradation is excessive water use. One of the most often used fibres in clothing manufacturing, cotton, needs a lot of water to grow. Cotton farming exacerbates water stress and can deplete freshwater supplies in areas where water is already scarce, such as portions of China and India.

Furthermore, the production processes of turning raw materials into textiles often rely on hazardous chemicals. For example, the dyeing and finishing of clothes uses harmful chemicals that contaminate streams and seriously endanger the health of surrounding residents and workers. Furthermore, the energy-intensive process of making textiles increases greenhouse gas emissions, exacerbating climate change. The worldwide transit of raw materials and completed goods further increases the industry's carbon footprint.

Moreover, the fast fashion model, characterized by rapid turnover of trends and cheaply made garments, has led to a culture of disposability. As consumers chase the latest styles at bargain prices, the lifespan of clothing has drastically shortened. Due to this "throwaway" mentality, mountains of textile waste are produced, most of which are dumped in landfills where it breaks down and emits the potent greenhouse gas methane. It can take decades or even centuries for clothing composed of natural fibres to decompose entirely.

The fashion industry's environmental impact extends beyond the production and disposal phases. The cultivation of raw materials often involves deforestation, particularly in the case of fabrics like rayon and viscose derived from wood pulp. Deforestation diminishes biodiversity, damages vital wildlife habitats, and accelerates climate change by lowering the planet's capacity to absorb carbon dioxide.

It's not just the environment that suffers from the fashion industry's practices. The social and ethical implications of fashion production are equally alarming. Many garments are made in sweatshops, where workers, often marginalized and vulnerable populations, toil in unsafe conditions for meagre wages. This exploitation not only violates human rights but also perpetuates cycles of poverty and inequality. By supporting ethical fashion, consumers can help break these cycles and promote a more just and equitable industry.

To address the fashion industry's environmental effects, changes must be made at every stage of the supply chain. This is a challenging task. Mitigating environmental harm requires the implementation of sustainable practices, such as employing organic or recycled materials, conserving energy and water, and developing closed-loop systems for textile recycling and reuse. However, the industry alone cannot bring about change. Consumers, through their purchasing decisions, can play a pivotal role. By choosing ethically produced and durable clothing, supporting brands that prioritize sustainability, and advocating for transparency and accountability within the industry, individuals can help shape a more environmentally conscious future for fashion.

Building a Sustainable Wardrobe

Building a sustainable wardrobe is not merely a trend but a conscious choice to reduce one's environmental footprint and promote ethical fashion practices. It involves adopting a mindset that values quality over quantity, longevity over fleeting trends, and transparency over ambiguity. Building a sustainable wardrobe is about redefining the relationship between consumers and clothing, shifting away from the fast fashion model of mass production and disposable fashion towards a more mindful and considered approach to dressing.

The cornerstone of a sustainable wardrobe is choosing clothes manufactured from eco-friendly materials and created under ethical working conditions. Choose natural and organic fibres like bamboo, hemp, linen, and organic cotton to lessen the impact on ecosystems and public health. These crops are also farmed without the use of hazardous pesticides or chemicals. To further reduce environmental harm, consider using materials like Tencel or modal made from renewable resources like wood pulp.

Equally important is considering the social and ethical aspects of clothing production. Supporting brands prioritizing fair labour practices, providing safe working conditions, and paying workers a living wage is essential for promoting social justice within the fashion industry. To ensure workers' rights are upheld throughout the supply chain, look for certifications such as Fair Trade or those from groups like the Ethical Trading Initiative.

One of the most important principles of sustainable wardrobe building is lifespan, in addition to labour and material considerations. Invest in classic items that are well-crafted and made to last the test of time rather than giving in to the temptation of quick fashion and its never-ending churn of trends. Better clothing has a longer lifespan and generally causes less environmental damage per wear than less expensive, throwaway clothing.

Embracing adaptability and variety is another part of creating a sustainable wardrobe. To maximize wearability and reduce the need for excessive consumption, choose clothing that can be combined and rearranged to make many ensembles. Investing in multipurpose products like reversible or convertible clothing or choosing classic silhouettes and neutral hues that go well with other items in your collection are some ways to achieve this.

Moreover, adopting a second-hand and vintage purchasing style is eco-friendly, supporting a circular economy, reducing waste, and adding distinctive pieces to your wardrobe. Thrifting lessens the overall environmental effect of fashion consumption by prolonging clothing life and minimizing the demand for new manufacturing. Furthermore, pre-owned and vintage apparel items frequently have backstories and tales to tell, giving your wardrobe personality and charm.

As consumers, we also have the power to influence change through our purchasing decisions and advocacy efforts. By supporting brands that prioritize sustainability and transparency and holding others accountable for their environmental and social practices, we can drive positive change within the fashion industry. Additionally, advocating for policies and regulations that promote sustainability and ethical practices, such as extended producer responsibility and transparency in supply chains, can help create a more sustainable future for fashion.

In conclusion, building a sustainable wardrobe is a personal choice and a collective responsibility to safeguard the planet and its inhabitants. By choosing carefully what we wear and how we consume it, we can decrease our environmental impact, promote moral behaviour, and encourage others to work with us to create a more just and sustainable fashion industry.

Thrift Shopping and Upcycling

Thrifting, once relegated to the realm of budget-conscious shoppers and vintage enthusiasts, has emerged as a cornerstone of sustainable fashion in recent years. Purchasing second-hand clothing offers an affordable alternative to fast fashion and contributes to reducing textile waste and promoting a circular economy. Customers may give second-hand clothing a new lease on life by thrift shopping, whether at traditional brick-and-mortar stores, online, or swap events. They can also find one-of-a-kind items that give their wardrobes personality and uniqueness.

One of its key benefits is that you may purchase items from second-hand stores to reduce waste. Millions of tons of clothing are in landfills yearly due to the fashion industry's astounding waste output. By shopping second-hand, consumers divert these garments from the waste stream and extend their lifespan, reducing the environmental impact of fashion consumption. Thrift shopping also helps alleviate the demand for new production, reducing the resources and energy required to manufacture new clothing items.

Moreover, thrift shopping fosters a culture of reuse and resourcefulness, encouraging consumers to value and appreciate the inherent qualities of pre-owned clothing. Instead of succumbing to the cycle of disposable fashion and constant trend turnover, thrifting promotes a more mindful approach to consumption, focusing on finding pieces that resonate with personal style and preferences rather than fleeting trends. This shift in mindset cultivates a more sustainable relationship with clothing and encourages creativity and self-expression through styling and outfit curation.

In addition to thrift shopping, upcycling has gained traction as a sustainable fashion practice that empowers individuals to transform old or discarded garments into new, unique creations. Unlike traditional recycling, which breaks down materials into their essential components for reuse, upcycling involves repurposing and reimagining existing materials to create something of higher value or quality. Upcycling allows for endless possibilities for creativity and self-expression, from turning old jeans into a denim skirt to transforming vintage scarves into statement tops.

One key principle of upcycling is minimizing waste through the use of existing materials. Upcycles reduce the demand for virgin resources and mitigate the environmental impact of fashion production by giving new life to old garments and textiles. Upcycling also encourages consumers to reconsider their relationship with clothing and to view garments not as disposable commodities but as valuable resources with potential for reinvention and reuse. The fashion industry's transition to a more circular model is consistent with the ideas of environmental stewardship and sustainability.

Furthermore, upcycling fosters a sense of empowerment and self-sufficiency, allowing individuals to take control of their wardrobes and express their creativity through DIY projects. Upcycling provides a platform for experimentation and innovation, regardless of skill level or experience, whether through simple alterations and repairs or more ambitious design transformations. This democratization of fashion creation challenges the traditional notions of consumerism and production, inviting individuals to actively participate in shaping their style narratives.

Apart from its advantages for the environment and creativity, thrift shopping and upcycling also preserve cultural heritage and promote social justice within the fashion industry. By supporting second-hand markets and independent upcycling initiatives, consumers can help preserve traditional craft techniques and support local artisans and entrepreneurs. Thrift shopping also provides an accessible and affordable option for individuals from diverse socioeconomic backgrounds to access quality clothing without perpetuating exploitative labour practices or contributing to environmental degradation.

In conclusion, thrift shopping and upcycling offer viable pathways towards a more sustainable and ethical fashion future. By embracing these practices, consumers can reduce their environmental footprint, support a circular economy, and foster creativity and self-expression. Thrift shopping and upcycling challenge the status quo of fast fashion and empower individuals to reclaim agency over their wardrobes and make conscious choices that align with their values and beliefs. Upcycling and thrift store shopping offer viable options that put people and the environment before profit, particularly in light of the fashion industry's struggles with waste, exploitation, and environmental damage.

Supporting Ethical Brands

Supporting ethical brands is one of the most critical steps in transforming the fashion industry into a more sustainable and socially conscious organization. Ethical companies are distinguished by their dedication to environmental stewardship, fair labour methods, and transparency in a time when quick fashion and mass production are prevalent. By endorsing these labels, buyers can use their purchasing power to advocate for a fashion industry that puts people and the environment before profit.

A distinguishing feature of ethical brands is their dedication to maintaining openness across the supply chain. Unlike conventional fashion brands that often operate behind closed doors, ethical brands are transparent about their sourcing, production methods, and labour practices. They provide consumers with insight into where and how their garments are made, allowing for informed purchasing decisions that align with sustainability and social responsibility values. This transparency fosters trust and accountability, enabling consumers to support brands that uphold high ethical standards.

Ethical businesses value fair labour practices and ensure that workers are treated with respect and dignity throughout the supply chain, in addition to transparency. This means paying living wages, providing safe working conditions, and respecting workers' rights to organize and collectively bargain. By supporting brands that prioritize ethical labor practices, consumers can help combat exploitation and create a more equitable fashion industry. Moreover, ethical brands often invest in community development initiatives and support local artisans and producers, empowering marginalized communities and promoting economic sustainability.

Environmental stewardship is another critical pillar of ethical brands, prioritizing sustainable materials, production methods, and packaging. From using organic and recycled fibres to implementing water-saving techniques and reducing carbon emissions, ethical brands strive to minimize their environmental footprint at every stage of the supply chain. They prioritize quality over quantity, producing garments designed to last and minimizing waste through responsible manufacturing practices. By supporting brands that prioritize environmental sustainability, consumers can help mitigate the negative impacts of fashion on the planet and support efforts towards a more circular and regenerative economy.

Furthermore, ethical brands often engage in advocacy and activism to address systemic issues within the fashion industry and promote positive change. Whether through campaigns for worker rights, calls for transparency and accountability, or initiatives to encourage sustainable practices, ethical brands use their platform to raise awareness and drive meaningful action. By supporting these brands, consumers can amplify their voices and contribute to broader efforts to transform the fashion industry into a force for good.

As consumers, we can shape the future of fashion through our purchasing decisions and advocacy efforts. By supporting ethical brands, we can send a powerful message to the industry that sustainability and social responsibility are non-negotiable. We can also force systemic change by holding brands responsible for their deeds and insisting on transparency, accountability, and moral behaviour. By doing this, we promote businesses that share our beliefs and help build a fashion sector that benefits people, the environment, and the coming generations.

To sum up, fostering an ethical and socially conscious fashion business depends on patronizing ethical brands. Leading by example, ethical brands encourage others to follow suit by prioritizing activism, environmental stewardship, fair labor standards, and transparency. We, the consumers, drive the need for ethical fashion and will continue shaping the industry's future. By endorsing brands that follow high ethical standards, we can establish a fashion ecosystem where people, the environment, and moral values come first.

CHAPTER VI

Eco-Friendly Parenting

Raising Children with Sustainable Values

In a society that is becoming more environmentally sensitive and linked, instilling sustainable values in children cannot be overstated. As the stewards of tomorrow, today's children will inherit the consequences of our actions and decisions regarding the planet's health and well-being. By raising children with sustainable values, parents and caregivers can shape future generations of environmentally conscious and socially responsible individuals who prioritize the planet's well-being and its inhabitants.

At the heart of raising children with sustainable values lies cultivating a deep respect and appreciation for the natural world. Children should be encouraged to explore and connect with nature early through outdoor play, nature walks, gardening, or wildlife observation. By fostering a sense of wonder and curiosity about the environment, parents can instil in their children a lifelong love and reverence for the natural world, laying the foundation for future environmental stewardship.

The cultivation of durable values in children is primarily dependent on education. To help kids understand how human actions are interconnected and affect the environment, parents can have age-appropriate conversations with their kids about environmental issues like pollution, climate change, and biodiversity loss. Children can be inspired to take action to safeguard the environment and develop a broader awareness of ecological concerns by being exposed to various viewpoints and experiences through books, movies, or hands-on learning activities.

Moreover, leading by example is essential in teaching children's sustainable values. Parents who model environmentally friendly behaviours, such as reducing waste, conserving energy, and making eco-conscious consumer choices, demonstrate to their children the practical importance of sustainability in everyday life. Parents can empower their children by incorporating kids' in-home practices like recycling, composting, and water conservation and inculcate a feeling of responsibility for positively impacting the environment. Encouraging children to participate in sustainable practices can also foster a sense of agency and efficacy.

Whether volunteering for environmental clean-up efforts, participating in community garden projects, or advocating for environmental causes, children can learn that their actions can make a difference in the world around them. Parents can nurture a sense of environmental responsibility and civic engagement from a young age by empowering children to take ownership of environmental issues and encouraging them to be active change agents.

Parents can help their children grow compassion and understanding for other living things by teaching them the value of animal welfare and ethical treatment of animals and taking practical steps in this regard. Through engaging in activities such as pet care duties, visiting animal sanctuaries, or having conversations about the environmental effects of animal husbandry, kids can gain a more profound comprehension of the interdependence of all living organisms and the moral implications of sustainable living.

Fostering critical thinking abilities and pushing youngsters to challenge prevailing narratives and social conventions are other aspects of instilling durable values in children. Parents may empower their children to make well-informed decisions consistent with their values and beliefs by teaching them to consider their choices' social, economic, and environmental effects. Fostering a critical thinking environment in kids about excessive consumerism, waste management, and social justice might motivate them to question the current quo and promote a more just and sustainable society.

In summary, instilling sustainable values in children is a complex process that involves empowerment, modelling, and instruction. Parents may encourage their children to take action, led by example, develop a solid connection to nature, and instil a lifelong commitment to sustainability and social responsibility. By doing this, they raise a generation of kind and responsible global citizens committed to building a better future for everybody and providing kids with the knowledge and abilities they need to navigate an increasingly complicated world.

Eco-Friendly Baby Products

The arrival of a new baby is a joyous occasion, but it also brings with it a host of decisions for parents, including choices about the products they use to care for their little one. In recent years, there has been a growing interest in eco-friendly baby products—items designed with the health of babies and the planet in mind. From diapers and clothing to toys and skincare, eco-friendly baby products allow parents to minimize their environmental footprint while providing safe and healthy options for their child's care.

Disposable diapers are one of the primary sources of environmental contamination in the baby care industry. Conventional disposable diapers can take hundreds of years to break down in landfills since they blend polymers, synthetic materials, and wood pulp. On the other hand, environmentally friendly diaper choices, like disposable diapers that degrade naturally or cloth diapers, provide a more sustainable option. Reusing cloth diapers several times helps reduce waste and save resources over time. Conversely, the environmental impact of biodegradable disposable diapers is diminished because they are composed of materials derived from plants, which decompose more quickly in landfills.

In addition to diapers, clothing is another area where parents can make eco-conscious choices for their babies. Organic cotton clothing, made from cotton grown without synthetic pesticides or fertilizers, is a popular choice for eco-minded parents. Organic cotton is better for the environment and gentler on babies' sensitive skin, as it reduces exposure to potentially harmful chemicals. Additionally, choosing clothing made from sustainable materials such as bamboo or hemp can minimize environmental impact, as these fibres require fewer resources to grow and process than conventional cotton.

When it comes to baby toys, parents can opt for products made from natural materials such as wood, organic cotton, or silicone. These materials are free from harmful chemicals and dyes, making them safer for babies to chew on and play with. Wooden toys are a sustainable option for families that care about the environment because they are strong, long-lasting, and biodegradable. Additionally, selecting toys that promote creativity and unrestricted play might lessen the need for excessive toy consumption and support a more sustainable and minimalist parenting style.

Skincare products are another area where there are many eco-friendly solutions for babies. Many traditional baby skincare products find preservatives, artificial perfumes, and other potentially dangerous substances. On the other hand, eco-friendly skincare products are made with natural and organic components that nourish and soothe a baby's sensitive skin. Seek out items devoid of artificial perfumes, phthalates, sulfates, and parabens. Also, choose brands that highly value transparency and sustainability in their manufacturing methods.

In addition to choosing individual eco-friendly baby products, parents can also take steps to reduce waste and conserve resources in their daily baby care routines. For example, using reusable cloth wipes instead of disposable wipes can significantly reduce waste and save money over time. Similarly, opting for multi-functional baby gear, such as convertible cribs and strollers, can extend their lifespan and reduce the need for additional purchases as babies grow.

In the end, choosing eco-friendly baby care options benefits both the environment and the health and well-being of infants. Parents can give their children a safe, healthy, and environmentally conscious start by choosing long-lasting products free of hazardous chemicals and created from sustainable materials. Furthermore, parents can ensure a better and more sustainable future for future generations by modelling eco-friendly behaviours and values and instilling a lifetime appreciation for sustainability and environmental stewardship in their kids.

Sustainable Education

Sustainable education is not merely about teaching environmental facts or instilling a sense of eco-consciousness; it is about fostering a deep understanding of the interconnectedness of human societies, economies, and ecosystems and enabling people to take an active role as change agents in a rapidly changing world. Sustainable education seeks to cultivate critical thinking, problem-solving skills, and a sense of responsibility towards the planet and its inhabitants, preparing students to navigate complex environmental and social challenges with resilience and creativity.

Central to sustainable education is recognizing that sustainability is a multifaceted concept encompassing environmental, social, and economic dimensions. Rather than treating these dimensions in isolation, sustainable education encourages students to explore their intersections and interdependencies, understanding that solutions to global challenges must address the complex interactions between human societies and the natural world. By implementing a comprehensive educational strategy, students acquire a more all-encompassing comprehension of sustainability and are more prepared to make significant and enduring contributions.

One of the fundamental principles of sustainable education is experiential learning, which emphasizes hands-on, real-world experiences that allow students to engage directly with environmental and social issues in their communities. Through experiential learning, students can apply their theoretical knowledge to real-world problems, leading to a deeper understanding of sustainability and a higher sense of personal agency. Examples of this type of learning include field trips to nearby ecosystems, service-learning projects with community partners, and extracurricular activities with a sustainability focus.

Moreover, sustainable education promotes interdisciplinary learning, encouraging students to draw connections between different subject areas and explore the complexities of sustainability from multiple perspectives. Students develop a more holistic understanding of sustainability by integrating concepts from science, social studies, mathematics, language arts, and other disciplines. They are better equipped to address complex, real-world problems that transcend traditional disciplinary boundaries. Interdisciplinary learning fosters creativity and innovation, as students are encouraged to think critically and explore unconventional solutions to sustainability challenges.

Sustainable education stresses the development of critical thinking and problem-solving abilities necessary for resolving sustainability concerns in a world that is changing quickly, in addition to experiential and multidisciplinary learning. Pupils are urged to challenge presumptions, assess evidence, and pose questions to gain the ability to understand complicated problems, pinpoint underlying causes, and come up with creative solutions. Sustainable education gives students the tools and perspective to deal with uncertainty and adjust to shifting social and environmental contexts by promoting a culture of inquiry and discovery.

Furthermore, sustainable education recognizes the importance of fostering a sense of empathy, compassion, and social responsibility in students, encouraging them to consider the needs and perspectives of others, both locally and globally. Through service-learning projects, community engagement initiatives, and cross-cultural exchanges, students develop a deeper understanding of the interconnectedness of human societies and the importance of solidarity and cooperation in addressing global challenges. By nurturing empathy and fostering a sense of global citizenship, sustainable education prepares students to become active and engaged members of their communities and advocates for social and environmental justice.

Ultimately, sustainable education is not just about imparting knowledge but about cultivating values, attitudes, and skills that empower individuals to live in harmony with each other and the natural world. By fostering critical thinking, interdisciplinary learning, experiential education, and a sense of empathy and social responsibility, sustainable education equips students with the tools and mindset needed to create a more sustainable and equitable future for all. Sustainable education provides a route to resilience, creativity, and constructive change as we face the pressing issues of social inequality, biodiversity loss, and climate change. This guarantees that the next generation is equipped to face the challenges of the twenty-first century with bravery and compassion.

CHAPTER VII

Green Workplaces

Creating Sustainable Office Environments

Creating sustainable office environments is essential to mitigate environmental effects and promote a sustainable culture inside workplaces. As the places where millions of people spend a significant portion of their daily lives, offices have a substantial environmental footprint. From energy consumption and waste generation to resource use and employee well-being, every aspect of an office environment presents opportunities for improvement. By implementing sustainable practices, businesses can reduce their ecological impact, enhance employee health and productivity, and contribute to broader environmental goals. The journey towards a sustainable office environment encompasses a holistic approach that includes energy efficiency, waste reduction, sustainable sourcing, green building design, and fostering a culture of sustainability among employees.

One of the main components of sustainable office spaces is energy efficiency. Office buildings are often energy-intensive, with heating, cooling, lighting, and electronic equipment accounting for significant energy use. One of the most effective strategies for reducing energy consumption is improving the energy efficiency of office buildings. This can be achieved through various measures, such as upgrading to energy-efficient lighting systems, implementing smart thermostats and HVAC systems, and ensuring proper insulation to minimize heating and cooling losses. Significant energy savings can be achieved with energy-efficient equipment and appliances, such as computers, printers, and refrigerators, accredited with the ENERGY STAR program.

Sustainable workplace spaces require not only energy efficiency but also the use of renewable energy sources. Businesses can lessen their dependency on fossil fuels and carbon footprint by making renewable energy investment alternatives like solar panels and wind turbines or acquiring green energy from nearby suppliers. Not only may on-site renewable energy generation lower carbon dioxide emissions, but it can also result in cost savings over the long run. Additionally, battery storage devices can improve energy resilience and provide a steady electricity supply during high demand or blackouts.

Waste reduction is another critical component of a sustainable office environment. Offices produce a lot of waste, mainly from single-use plastics, paper, and electronic waste. It is possible to dramatically lessen the environmental effect of office activities by implementing comprehensive waste management strategies prioritizing reduction, reuse, and recycling. Easy steps like establishing recycling centers, offering waste receptacles with labels identifying the contents, and promoting reusable products like water bottles, coffee cups, and cutlery can significantly impact. Businesses can also embrace digital solutions, such as digital signatures, electronic document management systems, and paperless invoicing and billing, to reduce reliance on paper documents.

Sustainable sourcing and procurement practices are essential for creating environmentally responsible office environments. This involves selecting products and services with a lower environmental impact, from office supplies and furniture to cleaning products and catering services. Selecting goods manufactured from sustainable or repurposed materials, opting for non-toxic and biodegradable cleaning agents, and supporting local and environmentally responsible vendors all reduce the ecological footprint of office operations. Furthermore, businesses can implement green procurement policies that set clear criteria for sustainability and require suppliers to adhere to environmental standards.

Green building design and construction principles are fundamental to creating sustainable office environments. Sustainable building designs may improve indoor air quality, create healthier and more productive workspaces, and drastically reduce energy and water use. Green buildings must use sustainable materials, energy-efficient HVAC systems, natural lighting and ventilation, and water-saving fixtures. Green building certifications offer guidelines and standards for creating and managing ecologically friendly structures. Examples of these include BREEAM (Building Research Establishment Environmental Assessment Method) and LEED (Leadership in Energy and Environmental Design).

Since indoor air quality directly impacts worker health and productivity, it is an essential component of sustainable workplace spaces. Inadequate ventilation, chemical cleaning agents, and off-gassing from furnishings and construction materials can contribute to poor indoor quality, leading to respiratory conditions, allergies, and diminished cognitive function, among other health problems. Offices can use strategies including boosting ventilation rates, utilizing air purifiers, using low-VOC (volatile organic compound) materials and furniture, and implementing green cleaning techniques to improve indoor air quality.

Water conservation is another essential element of a sustainable office environment. Offices can implement water-saving measures such as installing low-flow faucets, toilets, and urinals, using drought-resistant landscaping, and implementing rainwater harvesting systems. Regular maintenance and monitoring of plumbing systems can also help detect and fix leaks promptly, preventing water wastage. Additionally, businesses can promote water conservation awareness among employees by providing information and resources on reducing water use in the office and at home.

Any project to create a sustainable office must first cultivate a culture of sustainability in the workplace. This entails including workers at all levels in sustainability initiatives, from upper management to front-line personnel, and fostering an atmosphere that rewards and encourages sustainable behavior. Companies can accomplish this by implementing various tactics, including creating sustainability committees or green teams, offering education and training on sustainability, and defining precise objectives. A strong culture of sustainability can also be developed by promoting employee involvement in sustainability programs, such as energy-saving contests, trash reduction challenges, and volunteer opportunities for environmental projects.

Moreover, businesses can support sustainable commuting options for their employees to reduce the environmental impact of transportation. Offering incentives for carpooling, biking, walking, or using public transportation can help decrease the number of single-occupancy vehicles on the road and reduce greenhouse gas emissions. Employees may be encouraged to use active commuting methods by providing amenities like showers, changing rooms, and bike racks. Businesses might also look into flexible work options that can lessen the impact of commuting on the environment, such as telecommuting or remote work.

Implementing sustainable IT practices is another crucial aspect of creating a sustainable office environment. Offices rely heavily on technology, and IT operations' energy consumption and electronic waste can be significant. Businesses can adopt sustainable IT practices such as optimizing server usage through virtualization, implementing energy-saving settings on computers and other devices, and extending the lifecycle of IT equipment through proper maintenance and refurbishment. Recycling and appropriate disposal of electronic trash are also necessary to salvage valuable resources and keep hazardous materials out of landfills.

Furthermore, fostering wellness and health in the workplace is essential to sustainability. A workforce in good health is more resilient, engaged, and productive. Businesses can support employee health and wellness by providing ergonomic workstations, promoting physical activity through on-site fitness facilities or wellness programs, offering healthy food options, and creating relaxation and stress reduction spaces. Access to natural light and views of nature and incorporating biophilic design elements such as indoor plants and natural materials can also enhance employee well-being and connection to the natural environment.

Finally, measuring and reporting sustainability performance is essential for continuous improvement and accountability. Businesses can use tools and frameworks such as sustainability audits, environmental management systems (EMS), and corporate social responsibility (CSR) reporting to track their progress and identify areas for improvement. By setting measurable targets and providing frequent performance reports, businesses can show stakeholders that they are committed to sustainability. This will also encourage continued efforts to build a more sustainable working environment.

In conclusion, creating sustainable office environments is a multifaceted and ongoing process that requires a comprehensive approach and the engagement of all stakeholders. Businesses may greatly minimize their environmental effect and help create a more sustainable future by concentrating on energy efficiency, waste reduction, sustainable procurement, green building design, and cultivating a culture of sustainability. Additionally, promoting employee health and wellness, supporting sustainable commuting options, and implementing sustainable IT practices further enhance the sustainability of office environments. Through continuous measurement and reporting, businesses can ensure their sustainability efforts are practical and drive ongoing improvement. By creating sustainable office environments, businesses benefit the planet and enhance employee well-being, productivity, and overall organizational resilience.

Telecommuting and Remote Work

The rise of telecommuting and remote work represents a fundamental shift in the modern workplace, driven by technological advances, evolving employee expectations, and a growing emphasis on sustainability. Telecommuting allows employees to work from home or any location outside the traditional office, and remote work, which enables fully or partially distributed teams, has reshaped how businesses operate. This shift significantly impacts environmental sustainability, work-life balance, productivity, and organizational dynamics.

Understanding these impacts is crucial for companies seeking to implement effective telecommuting and remote work policies that benefit employees and the environment.

One of the most immediate and visible benefits of telecommuting and remote work is reducing commuting-related carbon emissions. Traditional commuting often involves personal vehicles, public transportation, and other forms of travel that contribute significantly to greenhouse gas emissions and air pollution. Businesses may dramatically reduce the number of cars on the road, cutting carbon emissions and improving air quality, by allowing workers to work from home. Research has indicated that working from home can lessen an individual's carbon footprint because it eliminates the need for daily trips, which can account for a significant amount of an employee's overall emissions.

In addition to reducing carbon emissions, telecommuting and remote work also decrease the demand for office space and associated resources. Traditional office environments require significant energy for heating, cooling, lighting, and powering office equipment. By shifting to remote work, companies can downsize their physical office space, lowering energy consumption and utility costs. Furthermore, reducing office space can decrease the need to construct and maintain office buildings, which also has a positive environmental impact. This shift can contribute to developing more sustainable urban environments, as reduced office space can lead to less urban sprawl and lower overall resource consumption.

Telecommuting and remote work can also promote a more sustainable use of resources within the workplace. Traditional offices often generate substantial waste, including paper, plastic, and electronic waste. Employees can adopt more sustainable practices by working from home, such as using digital documents instead of printing, recycling more effectively, and reducing disposable office supplies. Additionally, remote work can encourage businesses to adopt cloud-based technologies and digital collaboration tools, reducing reliance on physical resources and promoting more efficient and sustainable work practices.

The flexibility afforded by telecommuting and remote work also has significant implications for employee well-being and work-life balance. Workers who work from home are more adept at juggling work and personal commitments, reducing stress and improving quality of life. Employees may be happier in their jobs and more inclined to stick with a company that provides this freedom since they may customize their work environment to fit their needs. Furthermore, telecommuting may reduce the need for protracted, stressful commutes, enhancing mental and physical health. Businesses may cultivate a more engaged and motivated workforce, improving output and performance by supporting a better work-life balance.

To ensure successful adoption, firms must address the obstacles posed by the shift to telecommuting and remote work. Keeping remote teams' effective communication and collaboration going is one of the fundamental issues. Remote workers could feel alone and cut off from their coworkers without in-person encounters. To solve this problem and encourage seamless teamwork and involvement, businesses can invest in digital communication and collaboration tools like project management software, video conferencing, and instant messaging. Frequent virtual meetings and team-building activities can also increase remote employees' sense of belonging to the organization.

Another challenge is ensuring remote workers can access the tools to perform their tasks successfully. This includes providing reliable internet connections, ergonomic workstations, and access to essential software and tools. Businesses can offer stipends or reimbursements for home office expenses to ensure employees have the necessary infrastructure to work productively from home. Furthermore, providing training and support for remote work technologies can help employees adapt to new working methods and maximize their productivity.

Data security and privacy are essential factors to consider when working remotely or telecommuting. Businesses must implement robust cybersecurity measures to protect against data breaches and cyberattacks since employees access company networks and sensitive data from many locations. This involves securing remote communications using encryption technology, multi-factor authentication, and virtual private networks (VPNs). Furthermore, consistent cybersecurity best practice training can assist staff members in identifying and addressing risks and safeguarding firm information and resources.

Performance management is another area that requires careful attention in a remote work environment. Traditional performance evaluation methods, which often rely on direct observation and in-person interactions, may not be suitable for remote teams. Businesses must develop new strategies for measuring and managing employee performance, focusing on outcomes and results rather than physical presence. This can involve utilizing analytics and performance indicators, defining precise objectives and deadlines, and giving remote employees frequent support and feedback. By implementing a results-oriented approach, businesses may guarantee remote workers stay accountable and productive.

Employee well-being and mental health are essential considerations in a remote work environment. Employee burnout and stress can result from a lack of physical separation between work and personal life since establishing boundaries and detaching from their work might be difficult. Companies should encourage employees to emphasize self-care, set regular work hours, and take breaks to foster a healthy work-life balance. Furthermore, providing wellness initiatives, counseling services, and other mental health resources can aid in stress management and employee well-being.

The implementation of telecommuting and remote work also affects leadership and organizational culture more broadly. Leaders must adjust to support and manage their teams in new ways while establishing a culture of trust, accountability, and autonomy. This entails establishing expectations, communicating clearly, and allowing staff members to own their work. To keep a safe and effective workplace, leaders must also take the initiative to address the issues of working remotely, such as resolving feelings of isolation and preserving team cohesion.

Telecommuting and remote work also present opportunities for businesses to enhance their sustainability efforts and corporate social responsibility initiatives. By mitigating the ecological consequences of office operations and transportation, corporations can augment more general sustainability objectives and exhibit their dedication to environmental stewardship. Additionally, remote work can support social and economic sustainability by providing more inclusive and flexible employment opportunities. This can include hiring talent from diverse geographic locations, supporting work-life balance for employees with caregiving responsibilities, and promoting accessibility for individuals with disabilities.

Adopting digital technologies and innovation has also accelerated due to the move towards telecommuting and remote work. Businesses have had to rapidly implement and scale digital solutions to support remote work, leading to increased investment in cloud computing, digital collaboration tools, and automation technologies. This digital transformation can drive efficiencies and innovation, enabling businesses to operate more effectively and respond to changing market conditions. Additionally, by lowering the need for physical resources and encouraging more effective and data-driven decision-making, the growing usage of digital technology can aid sustainability initiatives.

While telecommuting and remote work offer numerous benefits, businesses must approach these changes thoughtfully and strategically. This includes conducting thorough assessments of their current operations, identifying potential challenges and opportunities, and developing comprehensive implementation plans. Policies about remote work can be made sure to be successful and to suit the needs of the workforce by involving employees in the process and asking for their opinions.

In conclusion, telecommuting and remote work represent a significant transformation in the modern workplace, offering numerous benefits for sustainability, employee well-being, and organizational efficiency. By reducing commuting-related carbon emissions, decreasing the demand for office space, and promoting more sustainable work practices, businesses can contribute to environmental sustainability and create a positive impact. Additionally, the flexibility and autonomy afforded by remote work can enhance employees' work-life balance, job satisfaction, and productivity. However, businesses must also address the challenges of remote work, including communication, resource access, data security, performance management, and employee well-being, to ensure successful implementation. Businesses may leverage the promise of telecommuting and remote work to build a more sustainable, resilient, and flexible workplace in the future by taking a deliberate and planned approach.

Promoting Sustainability in Business

Promoting sustainability in business has become an essential objective for organizations worldwide as they seek to mitigate environmental impacts, meet regulatory requirements, and respond to the growing demands of consumers, investors, and other stakeholders. Sustainability in business encompasses a broad range of procedures designed to lessen the environmental impact of operations, enhance social responsibility, and ensure economic viability. Achieving sustainability requires a comprehensive approach that integrates environmental, social, and financial considerations into all business operations and decision-making aspects. This section explores vital business sustainability strategies, including sustainable resource management, reducing greenhouse gas emissions, waste reduction and recycling, sustainable supply chains, sustainable product design, corporate social responsibility, stakeholder engagement, and adopting sustainability frameworks and certifications.

One of the foundational elements of promoting sustainability in business is sustainable resource management. Part of this energy and raw materials is the use of natural resources, such as water, to lessen their adverse environmental consequences and ensure their continued availability for future generations. Enterprises can implement diverse approaches to sustainable resource management, including enhancing energy efficacy, diminishing water usage, and ethically procuring materials. For example, adopting energy- saving technologies, such as intelligent building management systems, LED lighting, and energy-efficient HVAC systems, can improve energy efficiency. Companies can also reduce their reliance on fossil fuels and greenhouse gas emissions by investing in renewable energy sources like solar, wind, and geothermal. Sustainable resource management can also benefit from water conservation techniques, including lowering water- intensive operations, putting in place water recycling systems, and installing low-flow fixtures.

Reducing greenhouse gas emissions is another requirement for promoting sustainability in business. Businesses may significantly mitigate the effects of climate change by putting policies in place to reduce their carbon footprint. This can entail implementing carbon offset programs, enhancing energy efficiency, moving to renewable energy sources, and lowering emissions using science-based targets. Companies also investigate cutting-edge approaches like carbon capture and storage technologies. To cut emissions. Since businesses can utilize frameworks like the Task Force on Climate-related Financial Disclosures (TCFD) to convey climate-related risks and opportunities to stakeholders, reporting and transparency are also essential.

Recycling and waste reduction are crucial elements of a sustainable business strategy. The generation of waste, whether solid, hazardous, or electronic, has significant environmental and health impacts. Businesses can implement comprehensive waste management programs that prioritize reducing, reusing, and recycling waste materials. This may entail establishing waste reduction goals, executing recycling programs for commodities like paper, plastic, glass, and metals, and conducting waste audits to find areas for improvement. Businesses should also look into prospects for industrial symbiosis, which reduces overall waste generation and promotes a circular economy by using waste products from one process as inputs for another. Other successful waste reduction techniques include minimizing single-use plastics and encouraging the use of compostable or biodegradable alternatives.

Sustainable supply chains are crucial for promoting sustainability in business. Supply chains often involve complex networks of suppliers, manufacturers, and distributors, each with environmental and social impacts. Companies must ensure that their suppliers follow social and environmental criteria, like ethical labour practices, decreased environmental impact, and responsible sourcing of raw materials, to establish sustainable supply chains. Supplier audits, sustainability evaluations, and the creation of supplier codes of behaviour can all help achieve this. Businesses can also work with suppliers to adopt sustainable practices, like cutting back on waste, using less energy and water, and using sustainable packaging. To ensure that products satisfy sustainability standards, companies can utilize technology like blockchain to track the origin and path of items along the supply chain. This makes transparency and traceability even more crucial.

Sustainable product design is another critical strategy for promoting sustainability in business. This means creating objects that follow a lifecycle, beginning with obtaining raw ingredients and concluding with manufacture, use, and disposal at the end of the product's useful life—in mind. Recyclability is encouraged, product longevity is increased, and environmental effect is reduced through sustainable product design. Businesses can adopt eco-design principles, such as using sustainable materials, reducing resource use, and designing for disassembly and recycling. Additionally, companies can explore the concept of product stewardship, where they take responsibility for the environmental impact of their products throughout their lifecycle. This can include implementing take-back programs, where consumers can return used products for recycling or refurbishment, and designing products for modularity and upgradability to extend their lifespan.

Sustainability in business is an essential component of encouraging corporate social responsibility (CSR). CSR involves firms taking responsibility for their social, environmental, and economic impacts and actively contributing to the well-being of society. CSR initiatives can encompass various activities, such as community engagement, philanthropy, ethical labor practices, and environmental stewardship. Businesses can develop CSR strategies that align with their core values and miss on and set measurable goals to track their progress. In addition to improving employee retention and happiness, involving staff members in CSR projects can help the company develop a sustainable culture. To show stakeholders how committed they are to ethical business practices, companies can use CSR reporting frameworks like the United Nations Global Compact or the Global Reporting Initiative (GRI) to share their sustainability performance.

The effectiveness of sustainability initiatives in business depends on stakeholder engagement. Stakeholders, including employees, customers, investors, suppliers, and local communities, are vested in the company's sustainability performance. Engaging stakeholders in sustainability efforts can provide valuable insights, build trust, and enhance the credibility of sustainability initiatives. Businesses can engage stakeholders through various means, such as surveys, focus groups, public consultations, and collaborative partnerships. Additionally, companies can establish stakeholder advisory panels or sustainability committees to provide ongoing input and guidance on sustainability strategies.

Transparent communication and regular reporting on sustainability performance are essential for maintaining stakeholder trust and demonstrating accountability.

Adopting sustainability frameworks and certifications can help business's structure and validate their sustainability efforts. There are numerous frameworks and certifications available that provide guidelines and standards for sustainable business practices. For instance, the International Organization for Standardization (ISO) offers social responsibility standards (ISO 26000) and environmental management systems (ISO 14001). A company's sustainability credentials can be improved by obtaining certifications like LEED (Leadership in Energy and Environmental Design) for green buildings, Fair Trade for ethical sourcing, and B Corp for overall social and environmental performance. These frameworks and certifications provide a structured approach to sustainability, enable benchmarking against industry standards, and improve transparency and accountability.

Businesses can use innovation and technology in addition to these tactics to promote sustainability. Technological innovations like artificial intelligence, the Internet of Things (IoT), and big data analytics present new possibilities for maximizing resource utilization, cutting emissions, and enhancing sustainability performance. Big data analytics can reveal trends in energy usage and point out areas where efficiency might be increased. AI may be used to cut waste and improve supply chain operations. Proactive management and maintenance are made possible by IoT technologies, which allow for real-time monitoring of resource usage and environmental conditions. Businesses should also consider using cutting-edge technology, such as blockchain, to verify the integrity of sustainability claims and enhance supply chain traceability and transparency.

Leadership and corporate governance play a crucial role in promoting sustainable ability in business. Strong leadership commitment to sustainability sets the tone for the entire organization and drives the integration of sustainability into business strategy and operations. Business leaders can demonstrate commitment by setting ambitious sustainability goals, allocating resources for sustainability initiatives, and embedding sustainability into the corporate culture. Corporate governance structures, such as sustainability committees or dedicated sustainability roles, can provide oversight and accountability for trainability efforts. Additionally, incorporating sustainability criteria into executive compensation and performance evaluations can incentivize leaders to prioritize sustainability and drive progress toward sustainability goals.

Employee engagement and education are also vital for promoting sustainable ability in business. Employees are often the frontline implementers of sustainability initiatives and can provide valuable ideas and feedback. Companies can engage employees in sustainability efforts through training and education programs, sustainability challenges, and recognition and rewards for sustainable behaviour. Creating opportunities for employees to participate in sustainability initiatives, such as volunteer programs, green teams, or innovation contests, can also foster a sense of ownership and commitment to sustainability. Additionally, promoting a culture of sustainability within the organization, where sustainable practices are encouraged and celebrated, can enhance employee morale and drive collective action towards sustainability goals.

Consumers' awareness and desire for sustainable goods and services also drive businesses to adopt more sustainable practices. Consumers are increasingly seeking environmentally friendly, ethically sourced, and socially responsible products. Companies can respond to this demand by offering sustainable products, providing transparent information about their sustainability practices, and engaging in marketing and communication efforts to highlight their commitment to sustainability. In addition to improving brand loyalty and reputation, developing strong connections with customers based on mutual respect and trust can give businesses a competitive edge.

Finally, businesses can collaborate with other organizations, industry groups, and governments to achieve sustainability. Collaborative efforts, such as industry alliances, public-private partnerships, and multi-stakeholder initiatives, can drive collective action toward common sustainability goals. For example, businesses can participate in industry initiatives to reduce greenhouse gas emissions, collaborate with NGOs to support community development projects or work with governments to develop and implement sustainability policies and regulations. Collaboration can also facilitate knowledge sharing, innovation, and the scaling of successful sustainability practices across industries and sectors.

In conclusion, promoting sustainability in business is a complex, dynamic process that calls for an all-encompassing, coordinated strategy. Companies can make significant strides towards sustainability by focusing on sustainable resource management, reducing greenhouse gas emissions, waste reduction and recycling, sustainable supply chains, sustainable product design, corporate social responsibility, stakeholder engagement, and adopting sustainability frameworks and certifications. Leveraging technology and innovation, demonstrating strong leadership and corporate governance, engaging employees and consumers, and collaborating with other organizations and stakeholders are also critical for advancing sustainability in business. As businesses navigate the complex and evolving landscape of sustainability, they can create positive environmental, social, and economic impacts, enhance their resilience and competitiveness, and contribute to a sustainable future for all.

Encouraging Eco-Conscious Practices Among Employees

Encouraging eco-conscious practices among employees is critical to building a sustainable business. By fostering a green workplace culture, organizations can significantly reduce their environmental impact, enhance employee engagement, and promote a positive corporate image. Eco-conscious practices involve a range of behaviours and initiatives that reduce resource consumption, minimize waste, and promote environmental stewardship. This section explores comprehensive strategies for encouraging eco-conscious practices among employees, including leadership commitment, employee education and training, sustainable workplace policies, incentivizing green behaviours, creating a supportive infrastructure, and engaging employees through communication and community-building activities.

Leadership commitment is fundamental to promoting eco-conscious practices among employees. When leaders demonstrate a solid commitment to sustainability, it sets a precedent and inspires employees to adopt similar values. Leaders can show their commitment by integrating sustainability into the company's mission and values, setting clear environmental goals, and communicating the importance of sustainability in business operations. For example, top executives can participate in and endorse sustainability initiatives, such as energy-saving campaigns or community clean-up events, to visibly support eco-conscious practices. Additionally, leaders can ensure that sustainability is prioritized in strategic decision-making and resource allocation, reinforcing the importance of environmental stewardship across the organization.

Employee education and training are crucial for empowering employees to adopt eco-conscious practices. Providing comprehensive education on environmental issues and the impact of workplace activities on the planet helps employees understand the significance of their actions. Organizations can offer training programs on energy conservation, waste reduction, and sustainable commuting options. Workshops, webinars, and online courses can educate employees on practical steps to reduce their environmental footprint at work and at home. Additionally, incorporating sustainability topics into onboarding programs ensures new employees know the company's commitment to sustainability from the outset. Continuous learning opportunities and access to resources, such as informational newsletters or an internal sustainability portal, can keep employees informed and engaged.

Sustainable workplace policies are another effective strategy for encouraging eco-conscious practices among employees. Policies that promote resource conservation and waste reduction create a framework for sustainable behaviour. For example, companies can establish policies prioritizing energy efficiency, such as setting temperature controls in office spaces, encouraging energy-efficient appliances, and implementing automatic lighting systems. Waste reduction policies can include guidelines for recycling, composting, and minimizing single-use plastics. Furthermore, companies can develop green procurement policies that favour environmentally friendly products and services, such as recycled paper, non-toxic cleaning supplies, and energy-efficient office equipment. Organizations create clear expectations and standards for sustainable behaviour by formalizing these practices into policies.

Incentivizing green behaviours can motivate employees to adopt eco-conscious practices. Rewards and recognition programs that celebrate sustainable actions can drive participation and foster a culture of sustainability. For example, companies can implement a green rewards program where employees earn points for eco-friendly actions, such as carpooling, using public transportation, or participating in a company-sponsored environmental event. Rewards like gift cards, more vacation days, or contributions to environmental charity can be obtained by exchanging these points. Programs for acknowledgment, such as employee of the month awards for sustainability champions, can help draw attention to individual and group accomplishments. Furthermore, gamification strategies, like sustainability contests or challenges, can make environmentally friendly behaviors enjoyable and exciting, promoting broad involvement.

Enabling environmentally responsible behaviors in the workplace requires the creation of a supportive infrastructure. This entails giving staff members the resources and tools to adopt sustainable practices quickly. For instance, setting up recycling stations and clearly labeling bins for different types of waste can facilitate proper waste segregation. Installing energy-efficient appliances and encouraging reusable items like coffee mugs and water bottles can reduce resource consumption. Additionally, companies can invest in technologies that support remote work and virtual meetings, reducing the necessity of travel and the carbon impact linked to commuting and business trips. Providing facilities for bike storage and showers can also encourage employees to cycle to work, promoting sustainable commuting options.

Engaging employees through communication and community-building activities can strengthen adopting eco-conscious practices. Regular communication about the company's sustainability goals, progress, and initiatives keeps employees informed and motivated. This can include updates in company newsletters, sustainability reports, and intranet posts. Employee opinions, ideas, and questions are welcome during town hall meetings and sustainability forums. Organizing a green team or sustainability committee with workers from different departments can also increase involvement. These teams can lead projects, plan activities, and represent sustainability within the company. Tree-planting, clean-up drives, and eco- friendly training are a few examples of community- building activities that can help employees feel more connected and responsible for one another.

Tracking advancement and upholding accountability requires measuring and disclosing sustainable performance. Key performance indicators (KPIs) relating to environmental impact allow firms to monitor their success and identify areas for improvement. Employees can see the observable results of their work when performance data, such as decreases in energy use, trash production, or carbon emissions, is routinely shared with them. Transparency in reporting also builds trust and reinforces the company's commitment to sustainability. Additionally, setting short-term and long-term sustainability targets can provide clear goals for employees to work towards, fostering a sense of purpose and motivation.

Another critical aspect of promoting eco-conscious practices is integrating sustainability into the corporate culture. Embedding sustainability into the organization's core values and daily operations ensures that eco-conscious practices become second nature to employees. This can be achieved by igniting sustainability with the company's mission, vision, and strategic objectives. Managers and leaders can foster an environment of responsibility and environmental stewardship by modeming sustainable conduct. Additionally, sustainability can be integrated into performance evaluations and development plans, ensuring that employees are recognized and rewarded for contributing to the company's sustainability goals.

Collaborating with external partners and stakeholders can enhance the effectiveness of sustainability initiatives. Companies can leverage additional resources, expertise, and support for their sustainability efforts by partnering with environmental organizations, government agencies, and industry groups. For example, collaboration with local ecological NGOs can allow employees to participate in community-based sustainability projects. Engaging with industry people through sustainability networks or alliances can facilitate knowledge sharing and dissemination of best practices. Additionally, partnerships with suppliers and customers can drive sustainability throughout the value chain, promoting a holistic approach to environmental responsibility.

Leveraging technology and innovation can also play a significant role in encouraging eco-conscious practices among employees. Digital tools and platforms can streamline sustainability initiatives and enhance employee engagement. For example, mobile apps that track and reward sustainable behaviours, such as energy savings or waste reduction, can provide real-time feedback and incentives. Cloud-based project management platforms and video conferencing are virtual collaboration solutions that reduce travel and paper-based procedures. Adopting innovative office technologies like IoT sensors and automation systems can optimize resource use and create a more sustainable work environment. Embracing technological innovations, such as renewable energy solutions and sustainable product design, can further advance the company's sustainability goals.

Encouraging eco-conscious practices also involves addressing the environmental impact of remote work and telecommuting. While remote work can reduce the carbon footprint associated with commuting, it also presents challenges related to energy consumption and resource use at home. Companies can support remote employees in adopting sustainable practices by providing guidelines and resources for home energy efficiency, such as tips for reducing electricity use and information on renewable energy options. Additionally, offering stipends or reimbursements for energy-efficient home office equipment can help remote workers minimize their environmental impact. Promoting virtual collaboration and reducing the need for travel can further enhance the sustainability of remote work arrangements.

Employee well-being and mental health are also essential considerations in promoting eco-conscious practices. A healthy and engaged workforce is likelier to participate in sustainability initiatives and adopt sustainable behaviours. Companies can support employee well-being by creating a healthy and sustainable work environment, offering wellness programs, and promoting work-life balance. For example, providing access to natural light, indoor plants, and ergonomic furniture can enhance the workplace environment and support employee health. Additionally, offering flexible work arrangements and promoting a culture of work-life balance can reduce stress and improve overall well-being, making employees more receptive to sustainability efforts.

Sustaining momentum in sustainability programs requires creativity and constant improvement. Organizations should periodically examine and evaluate their sustainability plans to pinpoint opportunities for improvement and creativity. This may entail conducting sustainability audits, comparing results to industry norms, and getting input from stakeholders and staff. Continuous improvement can also be fueled by keeping up with the latest sustainability and best practices developments. Fostering an innovative work environment where staff members are free to propose and try out novel sustainability concepts can result in innovative fixes and improvements in environmental care.

In conclusion, encouraging eco-conscious practices among employees is a complex and dynamic process that calls for an extensive and integrated approach. Leadership commitment, employee education, and training, sustainable workplace policies, incentivizing green behaviours, creating a supportive infrastructure, and engaging employees through communication and community-building activities are critical strategies for fostering a green workplace culture. Companies can create an environment where eco-conscious practices thrive by embedding sustainability into the corporate culture, leveraging technology and innovation, supporting remote work, and prioritizing employee well-being. Collaboration with external partners and stakeholders, continuous improvement, and transparent reporting further enhance the effectiveness of sustainability initiatives. Businesses may benefit the environment, society, and economy, contribute to a sustainable future, and develop a resilient and engaged workforce by navigating the opportunities and challenges of encouraging eco-conscious practices.

CHAPTER VIII

Sustainable Gardening and Landscaping

The Benefits of Sustainable Gardening

Sustainable gardening is a practice that cultivates a rich and diverse ecosystem and contributes significantly to the well-being of individuals and communities. This gardening method emphasizes environmental stewardship, resource conservation, and biodiversity, fostering a harmonious relationship between humans and nature. Sustainable gardening involves using eco- friendly techniques and materials, conserving water, promoting soil health, and supporting local wildlife. The benefits of sustainable gardening are multifaceted, ranging from environmental and economic advantages to social and health benefits. This thorough investigation explores the advantages of sustainable gardening, highlighting its impact on ecosystems, climate change mitigation, resource conservation, biodiversity enhancement, community well-being, mental and physical health, and economic sustainability.

The sound effects of sustainable gardening are among the main advantages of it on ecosystems. Sustainable gardening reduces soil, air, and water pollution by using organic methods and avoiding synthetic chemicals. This practice enhances soil health, crucial for plant growth and maintaining a balanced ecosystem. Composting and crop rotation with cover crops are two organic gardening practices that replenish the soil with essential minerals while enhancing its fertility and structure. Healthy soil supports robust plant growth and helps retain water, reducing the need for irrigation. Additionally, sustainable gardening practices promote the proliferation of beneficial microorganisms and insects, creating a vibrant and resilient ecosystem.

Sustainable gardening also plays a significant role in mitigating climate change. Plants absorb carbon dioxide (CO_2) from the atmosphere during photosynthesis, acting as natural carbon sinks. Sustainable gardening lowers greenhouse gas emissions and helps trap carbon by increasing the number of trees and other vegetation planted. Furthermore, utilizing sustainable gardening techniques reduces the carbon footprint of traditional gardening methods, like applying and producing chemical pesticides and fertilizers, which need a lot of energy. A healthier and more stable climate is facilitated by sustainable gardening, which minimizes the use of fossil fuels and reduces emissions.

Water conservation is another critical benefit of sustainable gardening. Conventional gardening often involves excessive water use, leading to the depletion of local water resources and increased water bills. Water conservation and waste reduction are achieved through sustainable gardening techniques like mulching, drip irrigation, and planting drought-tolerant plants. Mulching, for instance, controls soil temperature, inhibits weed growth, and helps keep soil moisture. With drip irrigation, water is delivered straight to the plant roots, reducing runoff and evaporation. By implementing these strategies, sustainable gardening guarantees the effective use of water resources and encourages gardens to withstand dry spells.

Sustainable gardening is based on the idea of enhancing biodiversity. The range of living things in a specific location, such as fungi, animals, plants, and microorganisms, is called biodiversity. A diversified ecology ensures the garden's long-term stability and health by being more resistant to pests and environmental disturbances. Sustainable gardening techniques that support biodiversity include using native plants, establishing wildlife habitats, and avoiding monocultures. Bees, birds, butterflies, and other pollinators are among the local wildlife that benefit from the food and shelter provided by native plants that have adapted to their surroundings. Creating habitats, such as ponds, birdhouses, and insect hotels, further encourages the presence of beneficial species. By fostering biodiversity, sustainable gardening contributes to local flora and fauna conservation and enhances the ecological balance.

Sustainable gardening also offers numerous social benefits, particularly in urban areas. Community gardens, a form of sustainable gardening, provide green spaces for residents to grow food and interact with nature. These gardens promote social cohesion, reduce stress, and improve mental well-being. They offer opportunities for community members to engage in physical activity, learn about sustainable practices, and develop a sense of ownership and pride in their local environment. Community gardens also serve as educational platforms where individuals can learn about organic gardening, composting, and water conservation. These communal areas promote a feeling of camaraderie and teamwork, reinforcing interpersonal relationships and elevating the standard of living.

The advantages of sustainable gardening for both physical and mental health are widely established. Strength, flexibility, and cardiovascular health can all be enhanced by gardening as a kind of physical activity. Digging, planting, weeding, and watering work different muscle groups and increase physical fitness. Furthermore, research demonstrates that gardening and outdoor activities might reduce stress, anxiety, and depression in individuals. A sense of direction and accomplishment is experienced when one tends to plants and watches them flourish. The sights, sounds, and scents of nature—all of which are part of gardening— positively impact one's mental health. In addition to promoting mindfulness and relaxation, gardening enables people to connect with nature and derive comfort from its beauty.

Economic sustainability is another significant benefit of sustainable gardening. Individuals can reduce their grocery bills and increase food security by growing their own food. Homegrown produce is often fresher, more nutritious, and free from harmful chemicals compared to store-bought alternatives. Sustainable gardening promotes local economies by supporting local nurseries, garden centres, and farmers' markets. The demand for organic seeds, plants, and gardening supplies creates economic opportunities for businesses prioritizing sustainable practices. Additionally, sustainable gardening reduces the costs associated with water use, energy consumption, and chemical inputs, making it a cost-effective alternative to conventional gardening.

Furthermore, sustainable gardening practices can be applied to larger agricultural systems, contributing to global food security and environmental sustainability. Agroecology, which integrates sustainable gardening principles into agricultural practices, emphasizes the importance of ecological balance, resource efficiency, and social equity. Agroecological practices, such as intercropping, agroforestry, and integrated pest management, enhance soil fertility, conserve water, and promote biodiversity. These practices can increase crop yields, improve climate change resilience, and reduce synthetic input dependency. Farmers can create a more just and sustainable food system using agroecological and sustainable gardening techniques.

Sustainable gardening also encourages environmental stewardship and responsibility towards the planet. By engaging in eco-friendly gardening practices, individuals become more aware of their environmental impact and are motivated to adopt sustainable behaviours in other aspects of their lives. This awareness can lead to broader changes, such as reducing energy consumption, minimizing waste, and supporting conservation efforts. Sustainable gardening catalyses environmental education and advocacy, inspiring individuals to take action to protect and preserve natural resources.

Moreover, sustainable gardening supports pollinator populations, which are essential for producing many plants and producing food crops. Because they facilitate fertilization, transmit pollen from one flower to another, and increase plant diversity, pollinators like bees, butterflies, and hummingbirds are essential to the environment. Unfortunately, declining pollinator populations result from habitat loss, pesticide use, and climate change. Pollinators can find safe havens in gardens if sustainable gardening techniques are used, such as planting flowers that attract them, avoiding pesticides, and providing nesting places. Sustainable gardening maintains gardens' health and productivity and helps stabilize the larger ecosystem by promoting pollinator numbers.

The aesthetic and recreational benefits of sustainable gardening should be noticed. A well-maintained garden provides a beautiful, tranquil space for relaxation, recreation, and inspiration. With their diverse plant species, vibrant colors, and varied textures, sustainable gardens offer a visually appealing environment that enhances the aesthetic value of homes and communities. These gardens can also serve as outdoor living spaces where individuals can enjoy leisure activities, host gatherings, and connect with nature. Designing, creating, and maintaining a sustainable garden can be fulfilling and creative, allowing individuals to express their artistic talents and preferences.

Sustainable gardening can also lessen the impact of infrastructure and human activity by reducing urban heat island effects—a phenomenon where urban areas get hotter than their rural surroundings. Green spaces, such as gardens, parks, and green roofs, help cool urban areas by providing shade, reducing heat absorption, and promoting evapotranspiration. Sustainable gardening practices, such as planting trees, creating green roofs, and installing green walls, can significantly reduce urban temperatures and improve air quality. These cooling effects enhance the comfort and liability of urban environments, reducing the need for air conditioning and lowering energy consumption.

Sustainable gardening encourages waste reduction and the preservation of natural resources in addition to these advantages. Composting organic waste, utilizing rainwater collecting systems, and recycling garden materials are some practices that help reduce resource consumption and the environmental impact of gardening. For instance, composting turns leftover food scraps and garden waste into nutrient-rich compost, improving soil quality and lessening the need for synthetic fertilizers. By collecting and storing rainwater for use in agriculture, rainwater harvesting systems reduce the need for municipal water sources. Reusing materials creatively, like building garden structures out of recycled wood or turning old containers into plant pots, prolongs their useful life and reduces waste.

The educational benefits of sustainable gardening extend beyond the individual gardener to the broader community. Schools, universities, and community organizations can use sustainable gardens as educational tools to teach students and the public about environmental science, biology, and sustainable practices. Hands-on learning experiences in gardens can enhance students' understanding of ecological principles, food production, and conservation. These learning opportunities help people develop a sense of responsibility for the environment and provide them with the knowledge they need to make sustainable decisions. Community outreach programs, such as gardening workshops, seminars, and tours, further disseminate knowledge and inspire collective action toward sustainability.

Sustainable gardening also contributes to the preservation of cultural heritage and traditional knowledge. Many indigenous and local communities have long practiced sustainable gardening techniques adapted to their specific environments and artistic practices. These conventional methods, such as companion planting, seed saving, and natural pest control, offer valuable insights into sustainable land management and biodiversity conservation. By preserving and promoting these practices, sustainable gardening helps maintain cultural heritage and supports the livelihoods and resilience of indigenous and local communities.

Finally, sustainable gardening fosters a deep connection to nature and a sense of environmental stewardship. Engaging in sustainable gardening allows individuals to observe the intricate relationships between plants, animals, and the environment. This connection fosters a greater appreciation for the natural world and a commitment to protecting it. Sustainable gardening encourages mindfulness and a slower pace of life, allowing individuals to reconnect with the rhythms of nature and find joy in the simple pleasures of gardening. This sense of connection and stewardship is essential for fostering a culture of sustainability and ensuring our planet's long-term health and well-being.

In conclusion, the benefits of sustainable gardening are extensive and far-reaching, encompassing environmental, social, health, and economic dimensions. By promoting healthy ecosystems, mitigating climate change, conserving water, enhancing biodiversity, and supporting pollinator populations, sustainable gardening contributes to environmental sustainability. The social advantages enhance living quality and promote environmental stewardship through community development, education, and cultural preservation. The value of sustainable gardening is further demonstrated by its benefits for both spiritual and physical well-being and the financial gains from growing food and using fewer resources. By adopting sustainable gardening techniques, people and communities help to create a more resilient, harmonious, and sustainable relationship with the natural environment. Sustainable gardening has the potential to spark a global movement towards environmental stewardship and sustainability via cooperation, innovation, and ongoing education.

Techniques for Eco-Friendly Gardening

Eco-friendly gardening, also known as sustainable or green gardening, encompasses a variety of practices aimed at reducing environmental impact while fostering a healthy and productive garden. These techniques emphasize resource conservation, biodiversity, soil health, and natural pest management, ultimately creating a balanced ecosystem that benefits the gardener and the environment. Implementing eco-friendly gardening methods can transform any garden into a sustainable green space, contributing to broader environmental sustainability efforts. This section explores several critical techniques for eco-friendly gardening, including soil health management, water conservation, plant selection, natural pest control, and waste reduction.

One of the foundational techniques for eco-friendly gardening is managing soil health through organic practices. A flourishing garden's foundation is healthy soil, which supplies vital nutrients, encourages plant development and aids in moisture retention. Gardeners can add organic matter to their soil, such as compost, aged manure, and leaf mould, to increase the health of their soil. Composting garden and kitchen scraps can improve soil quality and minimize the quantity of organic waste in landfills. Mulching is another helpful method that entails adding a covering of living things, such as grass clippings, straw, or wood chips, to the soil. As mulch breaks down, it helps keep the soil moist, inhibits weed growth, and progressively strengthens the soil's structure. Avoiding industrial pesticides and fertilizers is essential because these substances might damage beneficial soil organisms and upset the ecosystem's natural equilibrium in gardens. Rather than hurting the ecosystem, natural amendments and organic fertilizers like fish emulsion, rock phosphate, and bone meal promote soil fertility.

Another essential component of eco-friendly gardening is water conservation. Using water wisely preserves a valuable resource, lowers the gardener's water expense, and promotes plant health. Installing a rainwater harvesting system, which gathers and saves rainwater for later use in the garden, is one efficient way to conserve water. Using barrels under downspouts to collect roof runoff can be a simple solution. Water is delivered directly to plant roots with drip irrigation and soaker hoses, which reduces evaporation and runoff.

Additionally, selecting drought-tolerant plants and grouping plants with similar water needs can reduce water usage. Creating a water-efficient garden layout, with zones designated for high, moderate, and low water use, helps manage water resources effectively. As was already indicated, adding organic mulch also significantly contributes to water conservation by lowering soil surface evaporation.

Plant selection is a crucial element of eco-friendly gardening, as choosing the right plants can enhance biodiversity, support local wildlife, and ensure a resilient garden. Since native plants are more suited to the local soil and temperature, need less watering and upkeep, and offer food and habitat for wildlife, they are frequently the best option for eco-friendly gardening. A diverse diversity of pollinators, including bees, butterflies, and hummingbirds, can be attracted and maintained by planting various plants with varying bloom dates. Growing specific plants next to one another that complement one another might improve plant health and lessen the need for chemical treatments. This practice is known as companion planting. For instance, putting marigolds next to veggies helps keep pests away, and growing legumes next to vegetables helps fix nitrogen in the soil, which benefits the plants nearby. Adding edible plants to the garden, like fruit trees, herbs, and vegetables, further lowers the ecological footprint of food transportation while supplying fresh produce.

Natural pest control is a fundamental technique in eco-friendly gardening, aiming to manage pests without harming the environment. Encouraging beneficial insects like ladybugs, lacewings, and predatory beetles can naturally reduce pest populations. Planting flowers that offer pollen and nectar, like fennel, dill, and yarrow, will draw these valuable insects to the garden. Installing insect hotels and providing habitat features like piles of rocks and logs can also support beneficial insect populations. Another natural pest control method uses physical barriers, such as row covers, netting, and collars, to protect plants from pests. Hand-picking pests, such as caterpillars and beetles, can be an effective and immediate solution for small infestations. Also, homemade pest repellents, such as garlic spray, neem oil, and soap solutions, can deter pests without using synthetic chemicals. Changing the site of crops every season is known as crop rotation, and it can interrupt the life cycles of pests and lower the risk of soil-borne illnesses.

Waste reduction and recycling are integral to eco-friendly gardening, promoting a closed-loop system where garden waste is reused and repurposed. Composting is the primary method for recycling organic waste, transforming kitchen scraps, garden trimmings, and leaves into nutrient-rich compost that improves soil health. Another helpful method is vermiculture, sometimes known as worm composting, which turns organic waste into valuable compost by employing worms to break it down. Grass cycling replenishes soil nutrients and lessens the demand for synthetic fertilizers by leaving grass clippings on the lawn following mowing. Reusing garden materials, such as repurposing old containers, using reclaimed wood for garden beds, and recycling plant pots, helps minimize waste and reduce the demand for new resources. Additionally, gardeners can reduce plastic use by choosing biodegradable or reusable products, such as fabric plant pots, wooden labels, and metal watering cans.

Enhancing biodiversity and maintaining local ecosystems through eco-friendly gardening also involves creating habitats for wildlife. By giving wildlife access to food, drink, shelter, and nesting places, you may draw a variety of beneficial species to your yard. Bird feeders, birdbaths, and nesting boxes can support bird populations, while shallow water dishes and dense plant cover can attract amphibians and reptiles. Planting various flowers, shrubs, and trees that produce berries, seeds, and nectar ensures a year-round food supply for wildlife. Creating a wildlife-friendly garden benefits the environment adds to the garden's aesthetic appeal, and provides opportunities for observing and enjoying nature.

Finally, sustainable garden design principles can enhance the functionality and sustainability of the garden. Designing with permaculture principles that mimic natural ecosystems can create a self-sustaining garden requiring minimal input and maintenance. Essential permaculture techniques include zoning (arranging plants based on their water and maintenance needs), creating guilds (plant communities that support each other), and implementing water-saving strategies such as swales and rain gardens. Integrating hardscape elements, such as permeable paving, rain gardens, and green roofs, into the garden design can manage stormwater runoff, reduce erosion, and enhance biodiversity. Using locally sourced and sustainable materials for garden structures, such as FSC-certified wood, recycled plastic, and natural stone, minimizes the environmental impact of garden construction.

In conclusion, eco-friendly gardening encompasses a range of techniques that promote environmental sustainability, resource conservation, and biodiversity. By focusing on soil health, water conservation, plant selection, natural pest control, waste reduction, and wildlife support, gardeners can create a thriving and resilient garden that benefits both people and the planet. Sustainable garden design principles further enhance the garden's functionality and ecological balance. As more individuals adopt eco-friendly gardening practices, the collective impact on the environment can contribute to a healthier, more sustainable future for all. Through education, innovation, and community engagement, eco-friendly gardening can inspire a broader movement towards environmental stewardship and sustainability.

CONCLUSION

In "Earth-Friendly Living: The Path to Sustainable Happiness - Embrace Eco-Conscious Habits for a Better Tomorrow," we have explored the multifaceted journey towards a sustainable lifestyle that balances ecological responsibility with personal well-being. This e-book has provided insights and practical advice on various aspects of eco-friendly living, from adopting sustainable gardening practices to supporting ethical brands and encouraging eco-conscious behaviours among employees. The main takeaway is this: every action, no matter how tiny, adds to the effort to protect our world for coming generations.

Sustainable living begins with awareness and a commitment to change. By making informed choices in our daily lives—reducing waste, conserving water, or selecting ethically sourced and environmentally friendly products—we can significantly reduce our ecological footprint. The sustainability principles extend beyond personal habits to influence broader societal norms and corporate practices. Essential elements in this process include pushing for laws that protect natural resources, encouraging companies to adopt sustainable business practices, and encouraging the use of renewable energy.

The benefits of embracing an eco-conscious lifestyle are profound. Not only does it lead to a healthier environment, but it also enhances our quality of life. Sustainable practices often result in cost savings, improved health, and a stronger sense of community. Gardening, for example, provides fresh produce and fosters a connection to nature and a sense of accomplishment. Similarly, supporting local and ethical businesses helps build resilient economies and promotes social equity.

Education and community engagement play pivotal roles in fostering a culture of sustainability. We can inspire others to take meaningful actions by sharing knowledge and resources. Community gardens, workshops, and public awareness campaigns effectively spread the message and build momentum towards a more sustainable future. Schools and other educational institutions greatly influence the attitudes of the next generation about the environment.

In conclusion, "Earth-Friendly Living: The Path to Sustainable Happiness" emphasizes that achieving sustainability is not an endpoint but a continuous learning, adapting, and growing process. Ensuring that our activities today maintain the ability of future generations to meet their requirements is about achieving harmony between our lives and the natural environment. We are paving the way for a brighter tomorrow where everyone can live happily and sustainably as we adopt eco-conscious behaviours. Following this route, we leave a legacy of resilience, environmental care, and hope for a vibrant planet.

Thank you for buying and reading/listening to our book. If you found this book useful/helpful please take a few minutes and leave a review on the platform where you purchased our book. Your feedback matters greatly to us.

www.ingramcontent.com/pod-product-compliance
Lightning Source LLC
Chambersburg PA
CBHW070248230526
45470CB00002B/518